HEALTHY WATER

HEALTHY PEOPLE

Healthy Water, Healthy People
Water Quality Educators Guide

Hach Scientific Foundation

2114 North Lincoln
Loveland, Colorado 80538

201 Culbertson Hall
Montana State University
Bozeman, Montana 59717-0575

International
WET
Water Education for Teachers

201 Culbertson Hall
Montana State University
Bozeman, Montana 59717-0575

Cover photo credits:
Clouds © Don Farrall
Sawtooth Mountains, ID © Dick Dietrich
Ocean Wave © Digital Vision
Proxy Falls, OR © Dick Dietrich
Rock Dwelling © John P. George

ISBN: 1-888631-12-0

First Printing 2003

The Watercourse is an award-winning nonprofit science, natural resources, and heritage education program and publisher located on the campus of Montana State University in Bozeman, Montana, U.S.A. The Watercourse has two major programs: Project WET and Project Archaeology.

International Project WET, a nonprofit water resources education program of The Watercourse and Montana State University, is located in Bozeman, Montana U.S.A. The mission of International Project WET is to reach children, parents, educators, and communities of the world with unbiased water education. Project WET's goal is to provide scientifically accurate and educationally sound water resources education materials, training courses, and networking services to water and education agencies/organizations for use in designing, developing, and implementing their own localized Project WET programs based on the Project WET model. International Project WET partners with agencies/organizations that share its mission, goal, and the following core beliefs:

- Water is important to all water users (e.g., business and industry, earth systems, energy, fish and wildlife, navigation/ transportation, recreation, rural and agricultural, and urban/municipal)
- Wise water management is crucial for providing tomorrow's children social and economic stability in a healthy environment
- Awareness of, and respect for, water resources can encourage a personal lifelong commitment of responsibility and positive community participation

Established in 1989, Project WET works with visionary sponsors, educators, resource professionals, business leaders, policy makers, and citizens in the creation, development, and implementation of their projects. Project WET responds to the needs of many diverse groups and relies on public and private partnerships to accomplish its work.

International Project WET reaches millions of people each year through their international delivery network. This extensive grassroots support is a hallmark of Project WET.

Word Usage, Grammar, and Writing Style
The writing style within this guide follows the Chicago Manual of Style, 14th edition; spelling is based on the Random House Unabridged Dictionary, 2nd edition. The term ground water is presented as two words based on the recommendation of the United States Geological Survey (USGS), the primary water management data agency for the country.

Printed on recycled paper.

Dedication

Dear Educator:

The primary support for Healthy Water, Healthy People comes from the Kathryn Hach Trust. Clifford Hach was dedicated to the science of quantitative analysis and developing tests and methods for the precise measurement of known constituents in water. Hach tests are complete, simplified, accurate and employed by water management professionals around the world.

There can be no life without water. The water management industry deeply probes analysis and fundamental characterization of water's properties.

Healthy Water, Healthy People will introduce fundamentals of scientific exploration, and work with the unique character of water chemistry and water behavior. The "good wishes" of this book are dedicated to the competent commitment of water management professionals and the dedication of Hach Company chemists, scientists and associates.

Kathryn Hach-Darrow
Co-Founder/Former Owner of Hach Company
Chairman
Hach Scientific Foundation

The mission of the Hach Scientific Foundation is to make evident the interdependence between science education and the public.

Preface

John Etgen, Director
Healthy Water, Healthy People Program

Water quality has been a priority topic of Project WET since the program's inception in 1985. Between 1991 and 1995, Project WET conducted a national curriculum-writing project with peer-nominated educators and water resources professionals representing all 50 states state and several territories. These outstanding educators were asked to submit a list of three of the most important water topics that Project WET should pursue to supplement the core water resources education guide. This national survey resulted in the identification of five priority topics:

- Watersheds and Rivers
- Wetlands
- Water History/Environmental History
- Ground Water
- **Water Quality, Environmental, and Public Health**

Project WET committed to raising funds from public and private sources to publish education materials and delivery programs on each topic. Healthy Water, Healthy People is a direct result of Project WET's national and international need for a water quality education component.

Why Healthy Water, Healthy People?

The purpose of this publication is to raise educators' awareness and understanding of water quality topics and issues by demonstrating the relationship of water quality to personal, public, and environmental health. This publication–especially when used in combination with the other Healthy Water, Healthy People materials —gives teachers, students, nonformal educators, water managers, treatment plant operators, and citizens an opportunity to explore water quality topics in an interactive, easy-to-use, hands-on format.

What is Healthy Water for Healthy People?

Healthy water is simply water that supports and sustains life. All living things use water. It allows our food to grow, the trees to transpire, and our bodies to perform. The quality of the water is affected by a variety of factors, both natural and human-related. How healthy this water must be depends on how it is used.

The United States Environmental Protection Agency employs a classification of "designated uses" to determine what level of health water must attain. Water quality standards, or allowable levels of contaminants, are assigned to each of these uses. The most stringent standards apply to drinking water used in public water supplies. Following that, in order, are fish and wildlife, recreation, agriculture and industry, navigation, and other uses (e.g., hydroelectric, marinas, ground water recharge,etc.).

Meeting these water quality standards requires testing and measurement of the contaminant levels, which in turn requires a broader understanding of the sources, interactions, and remedies for these contaminants. Healthy Water, Healthy People publications and materials illustrate and promote the broad concepts of water quality testing and monitoring for educational purposes, while encouraging those who wish to become more involved to contact their local water quality monitoring program leader or state water quality specialist.

Our Goal

The goal of Healthy Water, Healthy People is to make the complex concepts of water quality relevant and meaningful for you and those you teach. Please let us know how we are doing.

Table of Contents

Table of Contents

Table of Contents

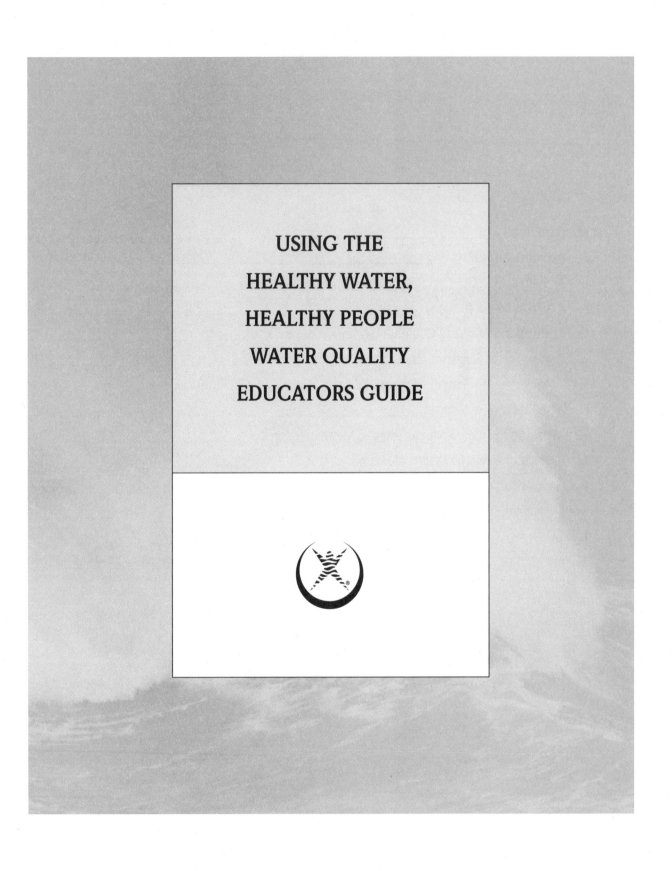

USING THE
HEALTHY WATER,
HEALTHY PEOPLE
WATER QUALITY
EDUCATORS GUIDE

Activity Format

Summary
A brief description of the concepts, skills, and affective dimensions of the activity.

Objectives
The qualities or skills students should possess after participating in the activity. NOTE: Learning objectives, rather than behavioral objectives, were established for activities. To measure student achievement, see Assessment.

Materials
Supplies needed to conduct the activity. (Describes how to prepare materials prior to engaging in the activity.)

Background
Relevant information about activity concepts or teaching strategies.

 ### HWHP Links
Provides internet links to additional information and resources about the concepts. Housed on the Healthy Water, Healthy People Web site www.healthywater.org. these links are updated periodically.

Procedure
Warm Up
Prepares everyone for the activity and introduces concepts to be addressed. Provides the instructor with preassessment strategies.

Grade Level
Suggests appropriate learning levels;
6-8 (Middle School)
9-12 (High School)

Subject Areas:
Disciplines to which the activity applies.

Duration:
Preparation time: The approximate time needed to prepare for the activity. NOTE: Estimates are based on first-time use. Preparation times for subsequent uses should be less.
Activity Time:
The approximate time needed to complete the activity.

Setting
Suggested site.

Skills:
Skills applied in the activity.

Vocabulary:
Significant terms defined in the glossary.

The Activity
Provides step-by-step directions to address concepts. NOTE: Some activities are organized into "parts." This divides extensive activities into logical segments. All or some of the parts may be used, depending on the objectives of instruction.

Wrap Up
Brings closure to the lesson and includes questions and activities to assess student learning.

Assessment
Presents diverse assessment strategies that relate to the objectives of the activity, noting the part of the activity during which each assessment occurs. Ideas for assessment opportunities that follow the activity are often suggested.

Extension
Provides additional activities for continued investigation into concepts addressed in the activity. Extensions can also be used for further assessment.

 ### Testing Kit Extensions
Provide additional activities for continued investigation of the activity concepts, specifically the Healthy Water, Healthy People Water Quality Testing Kits.

Resources
Lists references providing additional background information. NOTE: This is a limited list. Several titles are suggested, but many other resources on similar topics will serve equally well.

Activities Listed Alphabetically with Summaries

Water Quality Units

The *Healthy Water, Healthy People Water Quality Educators Guide* is ideal for teaching complete water quality units, as well as a supplement to existing curricula. This chart contains the suggested sequence of activities for teaching water quality units.

Unit Topics	Suggested Sequence of Activities
Water Quality Monitoring	Mapping It Out: Concept Mapping for Water Quality (6-12) Pg. 6 Hitting the Mark (6-12) Pg. 49 Grab a Gram (6-12) Pg. 29 Multiple Perspectives (6-12) Pg. 55 A Snapshot in Time (6-12) Pg. 61 Water Quality Monitoring: From Design to Data (9-12) Pg. 70
General Science Principles	Mapping It Out: Concept Mapping for Water Quality(6-12) Pg. 6 Hitting the Mark (6-12) Pg. 49 It's Clear to Me! (6-12) Pg. 10 Pollution-Take It or Leave It (6-8) Pg. 21 Grab a Gram (6-12) Pg. 29 Stone Soup (6-12) Pg. 35 Carts and Horses (6-12) Pg. 42 Going Underground (6-12) Pg. 187
Drinking Water, Healthy People and Communities	Mapping It Out: Concept Mapping for Water Quality (6-12) Pg. 6 A Tangled Web: Conducting Internet Research (6-12) Pg. 1 Pollution-Take It or Leave It (6-8) Pg. 21 Multiple Perspectives (6-12) Pg. 55 Footprints on the Sand (6-12) Pg. 90 Life and Death Situation (6-12) Pg. 125 Looks Aren't Everything (6-8) Pg. 99 Wash It Away (6-8) Pg. 121 Washing Water (6-12) Pg. 145 Picking Up the Pieces (6-12) Pg. 182 Setting the Standards (9-12) Pg. 107
Internet-based Research and Activities	A Tangled Web: Conducting Internet Research (6-12) Pg. 1 Life and Death Situation (6-12) Pg. 125 Multiple Perspectives (6-12) Pg. 55 Picking Up the Pieces (6-12) Pg. 182
Science Projects/ Scientific Method	A Tangled Web: Conducting Internet Research (6-12) Pg. 1 Carts and Horses (6-12) Pg. 42
Aquatic Ecology	Mapping It Out: Concept Mapping for Water Quality (6-12) Pg. 6 Turbidity or Not Turbidity (6-8) Pg. 83 Invertebrates as Indicators (6-8) Pg. 174 Water Quality Windows (6-8) Pg. 164 Benthic Bugs and Bioassessment (6-12) Pg. 154
Restoration	Mapping It Out: Concept Mapping for Water Quality (6-12) Pg. 6 Invertebrates as Indicators (6-8) Pg. 174 Picking Up the Pieces (6-12) Pg. 182 Going Underground (6-12) Pg. 187
Nonpoint Source Pollution	Mapping It Out: Concept Mapping for Water Quality (6-12) Pg. 6 Pollution-Take It or Leave It (6-8) Pg. 21 Footprints on the Sand (6-12) Pg. 90 Turbidity or Not Turbidity (6-8) Pg. 83 There Is No Point to this Pollution! (6-12) Pg. 136
Water Management	Mapping It Out: Concept Mapping for Water Quality (6-12) Pg. 6 Pollution-Take It or Leave It (6-8) Pg. 21 Hitting the Mark (6-12) Pg. 49 Multiple Perspectives (6-12) Pg. 55 There is No Point to This Pollution! (6-12) Pg. 136 Washing Water (6-12) Pg. 145 Picking up the Pieces (6-12) Pg. 182

How the Materials Work Together

The Healthy Water, Healthy People Program offers innovative, easy-to-use materials designed to make complex water quality concepts understandable and relevant for both teachers and students. The testing kits and publications were developed by teachers working with water quality experts, and appeal to beginning and advanced educators alike.

The Healthy *Water, Healthy People Water Quality Educators Guide* forms the foundation of the program, and contains twenty-five hands-on activities covering diverse water quality topics. *The Educators Guide* connects directly with the testing kits, encouraging students to explore the concepts even further.

The Healthy Water, Healthy People Testing Kits are designed to appeal to diverse learning styles and skill levels, depending on the chosen testing method.

The Healthy Water, Healthy People Testing Kit Manual serves as a technical reference text for both the testing kits and the guide, offering chapter overviews for eleven common parameters.

The Web site—www.healthywater.org—interacts with all of these components, serving as a clearinghouse of additional information offering links to case studies, activity aids (graphs, charts), and illustrations.

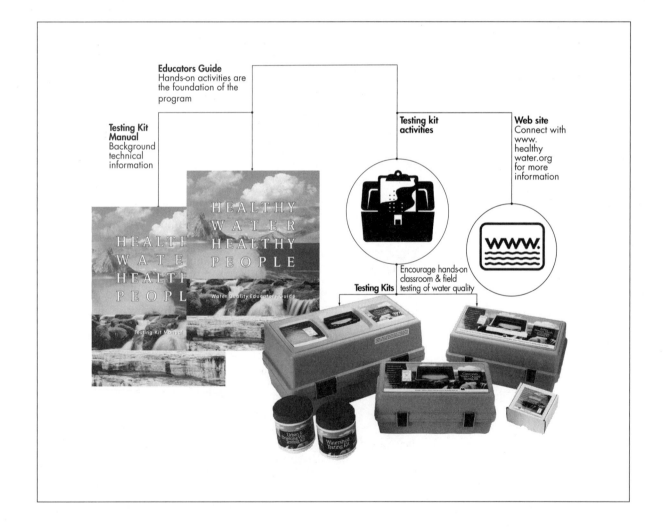

An Example of How the Publications and Testing Kits Work Together

The *Healthy Water, Healthy People Water Quality Educators Guide* contains twenty-five hands-on activities covering broad and specific water quality topics. A supporting publication, the *Healthy Water, Healthy People Testing Kit Manual,* serves as a technical reference covering eleven common water quality parameters. *The Healthy Water, Healthy People Testing Kits* serve as hands-on extensions to both of these publications, allowing for accurate testing of water quality parameters.

Challenge I: Classroom Teachers

An educator wants to teach her students about nonpoint source pollution, nutrient loading, and the resulting eutrophication.

The Solution:

Using Healthy Water, Healthy People Materials:

1. The instructor begins the unit by conducting "Mapping It Out. Water Quality Concept Mapping" activity from the *Healthy Water, Healthy People Water Quality Educators Guide.* Through this activity, the students build a concept map using their prior knowledge of nonpoint source pollution, nutrient loading, and eutrophication.

2. She then conducts the interactive activity, "There Is No Point To This Pollution!" from the *Healthy Water, Healthy People Water Quality Educators Guide.* This data analysis and mapping activity illustrates how nonpoint source pollutants and nutrients can enter waterways and accumulate, causing eutrophication.

3. A student then asks how phosphates are related to phosphorus, so she consults the "Phosphate" chapter in the *Healthy Water, Healthy People Testing Kit Manual* for the answer (or asks the student to search for it).

4. She then instructs her students to conduct a phosphate study of their local lake using a *Healthy Water, Healthy People Testing Kit* to test the phosphate levels.

5. She then conducts a demonstration of how nutrients move through the soil using the "Testing Kit Activities" section of the *Healthy Water, Healthy People Testing Kit Manual.*

6. She then assigns her students to conduct online research originating with the Healthy Water, Healthy People Web site–www.healthywater.org.

7. She concludes by revisiting the original concept map from the start of the unit. The students add to the map while the instructor assesses the knowledge gained between the beginning and the end of the unit.

Challenge II: Watershed Monitoring Project Leaders

The leaders of a water quality monitoring group want their group to understand the WHY and HOW behind water quality monitoring of their local waterway.

The Solution:

Using Healthy Water, Healthy People Materials:

1. The instructors begin the unit by conducting "A Snapshot in Time" activity from the *Healthy Water, Healthy People Water Quality Educators Guide.* This activity gives the students a foundation in understanding watersheds, data collection, and analysis.

2. They then conduct the interactive activity, "Hitting the Mark" from the *Healthy Water, Healthy People Water Quality Educators Guide.* This accuracy and precision activity gives the students an introduction into the importance of these concepts in relation to data collection.

3. A third activity, "Water Quality Monitoring: From Design to Data" is conducted by the instructors and the students are introduced to the fundamentals of study design in relation to a water quality monitoring project. The students also learn to analyze and compare real data.

4. A student then asks what can affect the dissolved oxygen content of a river, so the instructors consult the "Dissolved Oxygen" chapter in the *Healthy Water, Healthy People Testing Kit Manual* for the answer (or asks the student to search for it).

5. The instructors then ask the students to design and conduct a water monitoring study of their local river using the *Healthy Water, Healthy People Rivers, Streams, Ponds, and Lakes Testing Kit.*

6. They then conduct a demonstration of how dissolved oxygen is consumed by using the "Testing Kit Activities" section of the *Healthy Water, Healthy People Testing Kit Manual.*

7. The instructors then ask the students to conduct online research originating with the Healthy Water, Healthy People Web site–www.healthywater.org.

Healthy Water, Healthy People Program Overview

Water Quality Education Program

The Hach Scientific Foundation and Project WET (Water Education for Teachers), drawing on over fifty years of success in water quality test kit manufacturing, service, and water education, have partnered to create Healthy Water, Healthy People—a new and innovative water quality education program. Healthy Water, Healthy People is associated with Project WET, and originated as a result of a growing need for information and education on water quality, not only in the United States, but also around the world. The Healthy Water, Healthy People program encourages deep investigation of water quality topics and issues through development of user-friendly materials that are appropriate for all levels of users—from beginner to advanced.

Healthy Water, Healthy People recognizes that clean water is important for all people, prosperous economies, and natural systems. Water education must play an important role in providing opportunities for all citizens to learn about water quality in ways that are relevant and meaningful. Understanding the relationship of healthy water to healthy people will be critical as we collectively work to develop solutions for addressing continued water quality challenges and opportunities.

Mission Statement

The mission of Healthy Water, Healthy People is to reach children, young adults, educators, parents and communities with water quality education.

Goal

The goal of the Healthy Water, Healthy People Program is to facilitate and promote the awareness, appreciation, knowledge, stewardship, and understanding of water quality topics and issues, and to make evident the interdependence between science education and the public.

Audience

Healthy Water, Healthy People is for anyone interested in learning and teaching about contemporary water quality education topics:

- Upper Elementary through Secondary School Teachers
- Science Methods, Science Education and Environmental Science Professors
- River and Lake Monitoring Program Leaders, Drinking Water and Waste Water Facility Operators and Educators, Land and Water Managers, Conservation District and Extension Agents, Urban Program Members, Health Care Educators and Providers, Scientists, and Policy Makers
- Citizens—anyone interested in water quality

The Program

Healthy Water, Healthy People is an innovative, contemporary, and comprehensive water quality education program designed for anyone interested in learning about or teaching water quality. Healthy Water, Healthy People includes publications, testing kits, training and professional development, networking and support services, and an awareness campaign.

Publications

Healthy Water, Healthy People Water Quality Educators Guide
This 248-page activity guide is for educators of students in grades six through university level. The purpose of this guide is to raise the awareness and understanding of water quality topics and issues and their relationship to personal, public, and environmental health. Healthy Water, Healthy People will help educators address science standards through interactive activities that interpret water quality concepts and promote diverse learning styles, with foundations in the scientific method. This guide contains twenty-five original activities that link priority water quality topics to real-life experiences of educators and students.

Healthy Water Healthy People Testing Kit Manual
This technical reference manual is an excellent companion text that supports all of the *Healthy Water, Healthy People* publications and materials. The purpose of this

manual is to serve as a technical reference for the Healthy Water, Healthy People Water Quality Educators Guide and the Healthy Water, Healthy People Testing Kits, yielding in-depth information about eleven water quality parameters. The manual answers questions about water quality testing using technical overviews, data interpretation guidelines, case studies, chemical formulas, testing kit activities, laboratory demonstrations, and much more.

Healthy Water Healthy People KIDS (Kids in Discovery Series) Activity Booklet

This colorful, fully illustrated 16-page activity booklet is for fourth through seventh grade students. Through informative text, activities, investigations, and experiments, the Healthy Water, Healthy People KIDS booklet is designed to illustrate water quality topics and issues and make them intriguing, relevant, and fun for kids.

Testing Kits

Developed in cooperation with the Hach Company, a worldwide leader in water quality testing equipment, these water quality testing kits include all materials and equipment needed for field and classroom analysis of water samples. The testing kits are cross-referenced with

the above publications and allow students to conduct in-depth investigations and analysis of water quality parameters and issues. Healthy Water, Healthy People Testing Kits are available for a variety of parameters, grade levels, skills, and prices. Each Healthy Water, Healthy People testing kit package includes the *Healthy Water, Healthy People Water Quality Educators Guide*, a Testing Kit, and the *Healthy Water, Healthy People Testing Kit Manual.* To order a Healthy Water, Healthy People Testing Kit Package, contact the Healthy Water, Healthy People headquarters toll free at 1-866-337-5486, order online at www.healthywater.org, or contact us by email at healthywater@montana.edu. For Testing Kit technical support, contact Healthy Water, Healthy People.

Healthy Water, Healthy People FirstStep™

Healthy Water, Healthy People Watershed Testing Kit

Healthy Water, Healthy People Drinking Water Testing Kit–Urban, Rural, School

Healthy Water, Healthy People Classroom Drinking Water Testing Kit

Healthy Water, Healthy People Rivers, Streams, Ponds, and Lakes Testing Kit

Healthy Water, Healthy People Advanced Water Quality Testing Kit

Healthy Water, Healthy People Macroinvertebrate Investigation Kit–MacroPac™

Training Seminars, Workshops, and Institutes

Healthy Water, Healthy People provides customized water quality training opportunities for educators, corporations, communities, government agencies, educational conferences, and non-governmental institutions. People and organizations interested in working with the Healthy Water, Healthy People program to sponsor training seminars, workshops, and professional development institutes for staff and constituencies should contact Healthy Water, Healthy People to discuss dates and locations. A national and international network of water quality education trainers will be established. The program will help people and organizations design, develop, and implement their own water quality education programs. To schedule training, contact the Healthy Water, Healthy People staff toll-free at 1-866-337-5486.

Networking and Support Services

Healthy Water, Healthy People is committed to assisting people in their water quality education efforts. Staff members have extensive experience in water quality education programming and can help education providers answer questions.

• Education support—through Healthy Water, Healthy People staff 1-866-337-5486

• Network Communications Support—Healthy Water, Healthy People Newsgroup, subscribe at www.healthywater.org. This rapidly developing international network includes scientists, water educators, classroom teachers, water monitoring program leaders, healthcare professionals, and others interested in contemporary water quality education topics and issues.

WE CARE!® Campaign

If you care about healthy water and healthy people and want to demonstrate this commitment while contributing to the public's understanding of this critically important topic, the WE CARE!® Campaign is for you. The WE CARE!® Campaign is a national and international water quality awareness initiative designed for public and private organizations and corporations to

highlight their efforts to manage, protect, and restore water resources. There are amazing efforts taking place all around the world to provide people with clean and abundant water supplies and to sustain natural systems. Healthy Water, Healthy People staff can help these organizations demonstrate their support for healthy water in creative and customized ways. Contact the Director for more information.

Sponsors

The Healthy Water, Healthy People program is sponsored by the Hach Scientific Foundation and Project WET. Project WET and Healthy Water, Healthy People are programs of the Watercourse based at Montana State University.

Hach Scientific Foundation

The mission of the Hach Scientific Foundation is to foster and support science and science education, and to make evident the interdependence between science education and the public.

Financial support for additional Healthy Water, Healthy People projects has been provided by the Nestlé Waters of North America, the Hach Company, and Project WET USA Network Sponsors.

Sponsorship and Contact Information

Individuals or organizations interested in sponsoring Healthy Water, Healthy People should contact:

Director
Healthy Water, Healthy People Program

Healthy Water, Healthy People
Montana State University
PO Box 170575
Bozeman, MT USA 59717-0570
Phone: 1-866-337-5486
Fax: 406 994-1919
healthywater@montana.edu
www.healthywater.org

Acknowledgments

Sponsors

The *Healthy Water, Healthy People Water Quality Educators Guide* was published through a partnership between the Hach Scientific Foundation and International Project WET. The mission of the Hach Scientific Foundation is to foster and support science and science education, and to make evident the interdependence between science education and the public.

Publication Team and Contributors

Publisher
Dennis L. Nelson

Project Manager
John E. Etgen

Primary Authors
John E. Etgen
Keri Garver

Contributing Authors and Researchers
Dennis L. Nelson
Bruce J. Hach
Ann DeSimone
Cassie Murray
Linda Hveem
Erynne Dues Joyner
Tawni Thayer
Savannah Barnes
Sandra DeYonge

Production Coordinators
John E. Etgen
Ivy Davis

Editors
Anne Taylor
John Richardson
Savannah Barnes

Designer
Ivy Davis, *Studio I.D.*

Logo Design
Ivy Davis, *Studio I.D.*

Financial Management
Stephanie Ouren

The Watercourse and International Project WET Staff

Dennis Nelson,
Executive Director, The Watercourse and International Project WET
Stephanie Ouren,
Associate Director and Chief Financial Officer, The Watercourse and International Project WET
Linda Hveem,
Assistant to the Executive Director, The Watercourse and International Project WET
Gary Cook, *Director*
Project WET U.S.A.
John Etgen, *Director*
Healthy Water, Healthy People
Bonnie Satchatello-Sawyer,
Director, Native Waters
Jeanne Moe, *Director*
Project Archaeology
Karen Filipovich, *Director*
The Montana Watercourse
Savannah Barnes, *Writer*
The Watercourse and International Project WET

Rab Cummings,
National WETnet Coordinator, Project WET U.S.A.
Kristi Neptun
Project WET Montana Coordinator
Teresa Cohn, *Associate Director Native Waters*
Scott Frazier,
Native Waters Program Coordinator
Beau Mitchell,
Native Waters Researcher
Elisabeth Howe,
Co-director, Discover a Watershed Series: the Colorado Project
Justin Howe,
Co-director, Discover a Watershed Series: the Colorado Project
Sally Unser, *Program Assistant*
Gro Lunde, *Accounting Assistant*
Christine Gmitro,
Administrative Assistant
Helga Pac,
Administrative Assistant
Michelle Lebeau,
Adult Education Coordinator, Montana Watercourse
Duncan Bullock, *Graphic Designer Native Waters*
Katie Barcus, *Student Researcher Native Waters*
Dusty Hirsh-Mitchell, *Student Researcher Native Waters*
Ann DeSimone, *Student Researcher Healthy Water, Healthy People*

Acknowledgments

Healthy Water, Healthy People would like to thank the following persons and organizations for their donation of photographs and support for this guide:
Katie Alvin, Gallatin County MT Conservation District; Alexandra Antonioli, International Science Fair Winner; G. Donald Bain, The GeoImages Project; Savannah Barnes; Carolyn Bell, United States Geological Survey; Chelsea Booth, Fisheries & Oceans Canada; Sean B. Brubaker, Hach Company; Joe & Monica Cook, Artfamily.com; Dr. Stephan Custer, Montana State University; Deron Davis, Petey Giroux, Georgia Project WET; Ivy Davis, Studio I.D.; Stephen Delaney, United States Environmental Protection Agency; Harmony Demaniow; Linda DeMink, Montana Science Fair, University of Montana; Ann J. DeSimone; Tom DuRant, Harper's Ferry Center, National Park Service; Richard Fields, *Outdoor Indiana Magazine;* Keri Garver; George A. Grant, National Park Service History Collection; Hach Scientific Foundation; Lyn Hartman, Hoosier RiverWatch; Carla Hoopes, Montana's Statewide Noxious Weed Awareness and Education Campaign;

Paul Hunt, Michigan State University; Linda Hveem; Cheryl Mathews and Karen McEneaney, Yellowstone National Park; Allen McNeal, Natural Resources Conservation Service; Cassie Murray; Michael Quinn, Grand Canyon National Park; Kristi Rose, Chattahoochee RiverKeepers; Brenda Rupli, National Oceanographic and Atmospheric Administration; Susan Schultz, Indiana Project WET; Marcy Seavey, Iowa Academy of Science; Pete Schade, Montana Watercourse; Jessica Sparger, Indiana Department of Natural Resources; Sam Tamburro, Cuyahoga National Valley National Park; Tawni Thayer; Fred Walden, Indiana Teacher; LaVonda Walton, United States Fish and Wildlife Service; Water Education Foundation; Michelle White, Gallatin Blue Water Task Force; Wendy Williams; Gallatin County MT Conservation District.

Thanks to those who contributed maps and graphics to this guide: Tony Thatcher, DTM Consulting, for the topographic maps developed for the activities: *A Snapshot in Time, Looks Aren't Everything, There Is No Point to this Pollution.*

Katie Alvin, Carto-Logic, GIS for the GIS maps in the activity: *Water Quality Monitoring: From Design to Data.*

Special thanks to the following individuals for their help with this guide:
Katherine Hach-Darrow, Bruce J. Hach, and Pat Smith, Hach Scientific Foundation; Dennis Nelson; Linda Hveem; Stephanie Ouren; Ivy Davis; Ann DeSimone; Tawni Thayer; Cassie Murray; Harmony Demaniow; John Niemoth; Rich Paco Mason; Sheila Eyler; Eileen Tramontana; Marcy Seavey; April Rust; Súsan Schultz; Patti Vathis; Ann Regn; Cindy Grove; Mary Ann McGarry; Amy Picotte; Joey Breaux; Jean May-Brett; Sandra DeYonge; Pete Schade; Tina Laidlaw; Michelle White; Melissa Aquino; Julie Cronk; Tom Aspelund; Denton Slovacek; Barbara Martin; Eric Umbreit; John Parsons; Patrick Bowman; Jim Lafley; Rita Vasquez; Guillermo Larios; Jukka Holopainen; and to Jeannie and Cooper for their unending support for the project director.

Writing Workshops Participants

Writing Workshop Participants

Special thanks to all of the educators and experts who participated in the writing workshops, and whose activity ideas form the backbone of this guide.

Project WET Coordinators:

Rob Beadel, Arkansas; Gary Cook, Project Wet USA; Janice Conner, South Carolina; Rab Cummings, Montana; Nancy Kimbro, Delaware; Sara Kneipp, Texas; Terry Lewis, South Dakota; Laurina Lyle, Tennessee; Judy Maben, California; Diane Maddox, Kansas; Pamela Mathis, N. Mariana Islands; Mary Ann McGarry, Maine; Susan McWilliams, Oregon; Pauline Nystrom, Canada; Mary Pardee, Wisconsin; Amy Picotte, Vermont; April Rust, Minnesota; Pete Schade, Montana; Susan Schultz, Indiana; Marcy Seavey, Iowa.

Special Guests:

Tom Aspelund, Hach Company, Colorado; Patrick Bowman, Hach Company, Colorado; Jean May-Brett, NSTA District Director/Environmental Educator, Louisiana; Dr. Phil Butterfield, Montana State University; Teresa Byers, Middle School Teacher, Kentucky; Sandra DeYonge, International Project WET; Rick Edelen, High School Teacher, Montana; John Etgen, Healthy Water, Healthy People; Susan Fowler, Environmental Educator, Indiana; Hach Scientific Foundation Board and Staff, Colorado; Bethany Hansen, Environmental Educator, Wyoming; Lyn Hartman, Water Monitoring Coordinator, Indiana; Dana Hudson, Agriculture Extension Educator, Vermont; Linda Hveem, International Project WET; Dr. Clain Jones, Montana State University; Roger Lampe, Middle School Teacher, Nebraska; Dr. Clayton Marlow, Montana State University; Barbara Martin, Hach Company, Colorado; Natalie McCuiston, Middle School Teacher, Washington DC; Dennis Nelson, International Project WET; Julie Popham, Middle School Teacher, South Dakota; Lynne Quinnett-Abbott, Environmental Educator, Canada; Billie Schaffer, Middle School Teacher, Washington DC; Stephanie Simmons, Middle School Teacher, Colorado; Denton Slovacek, Hach Company, Colorado; Dr. Debra Thrall, University Professor, New Mexico; Briana Timmerman, Environmental Educator, South Carolina; Eric Umbreit, Hach Company, Colorado; Terry Zimmerman, High School Teacher, Florida.

International Fieldtest and Review

The activities in this guide were reviewed, tested, and retested by teachers, informal educators, water quality experts, university professors, and other interested parties from three countries and thirty-two states. Special thanks to all of these participants, who included: • 103 classroom teachers and over 3,000 students • 156 environmental educators • 76 content experts

Healthy Water, Healthy People Fieldtesters and Reviewers:

Alabama
Steve Henderson
Arizona:
Kerry Schwartz
Arkansas:
Amy Blackard
Ann Jones
California:
Judy Knott
Colorado:
Russell Clayshulte
Judith Daley
Steve Frank
Kevin Gertig
Hach Scientific
Foundation Staff
Barb Horn
Tina Laidlaw
Loretta Lohman
Pam Pickle
Sarah Rogers
Eric Umbreit
Florida:
Elise Cassie
Karen Hamilton
Michella Millington
Liberta Scotto
Eileen Tramontana
Georgia:
Deron Davis
Charles Richardson
Illinois:
Kathy Budach
Augustyn
Toni French
Susan Grabowski
Harry Hendrickson
Larry McPheron
Cathy Murges
Chris Parson
Stephanie Smith
Carol Widegren
Indiana:
Brooke Artley
Jeff Beck
Bill Brenneman
Amanda Burk

Amy Carpenter
Sue Crafton
Barbara Cummings
Lyn Hartman
Douglas Johnson
Pamela Katsimpalis
Greg Kiel
Mark Koschmann
Debra Lefever
Stacy McIntyre
Marquita Manley
Jeffrey Martin
Judy Morran
Daniel Robinson
Susan Schultz
Raymond Shepard
Lynn Stevens
Lenore Tedesco
Elma Thiele
D Angie Tilton
Lynn Wilhelm
Lana Zimmer
Iowa:
Mohammad Iqbal
Bob Libra
Marcy Seavey
Kansas:
Beth Carreno
Alison Louise Reber
Kentucky:
Melissa Dieckmann
Maine:
Mary Gilbertson
Maryland:
Elena Takaki
Massachusetts:
Barbara Eddy
Jim Lafley
Michigan:
Shawn Len
Mary Markham
Dave Rowe
Janet Vail
Diana Vermeulen
Minnesota:
Kamal Alsharif
Richard Cairn

Larry Dolphin
Kathy Dummer
Terry Dybsetter
Angie Becker Kudelka
Sue Jessee
Patty Riley
April Rust
Bonnie Schnobrich
Andrea Lorek Strauss
Mark Zembryki
Montana:
Rod Benson
Allen Bone
Joe Bradshaw
Phillip Butterfield
Jennifer DuBois
Clain Jones
Kristi Kline
Janet Kowles
Gail Miller
Colleen Osborne
T. Toller
Bill Tramp
Eric Wilson
Nebraska:
McKenzie Barry
Harry Heafer
Roger Lampe
Mike Sarchet
New Hampshire:
Karen Applegate
Lise Bofinger
Nicole Clegg
Nancy Large
Thomas Liveston
New Mexico:
Bryan Swain
Cheri Vogel
North Dakota:
Kim Belgarde
Archie Gronvold
Sherry Heilmann
CaraLee K-Heiser
Ila LaChapelle
Ohio:
Carolyn Cheek
Andrea Davidson

Jeannie Detmer
Mary June Emerson
Joyce Gottron
Daniel Kush
Lora Meredith
Vicki Morrow
Heather Moser
Amanda Podach
Trisha Schroeder
Dawn Wrench
Pennsylvania:
Laurel Bell
Robert Born
Sandra Cooley
Pam Diesel
Barbara Fauset
Vicki Fell-Pleier
Janis Gadpw
Antonia Giancola
Nancy Guidotti
Irene Guthrie
Ellen Ibarra
David Jarvie
Gary Lee
Heidi Lucas
Janet Masser
Patricia McGinnis
Laura McIntire
Jane Monaghan
Brenda Seigh
John Spagnuolo
Esther Ulery
Patricia Vathis
South Dakota:
Steve Jackson
Texas:
Bill Doucette
Vinay Dulip
Stanley Frazer
Christine Hawthorne
Kodi Jeffery
Adah Stock
Utah:
Andree' Walker
Vermont:
Gay Craig
Amy Picotte

Virginia:
Carrie Almli
Patrick Fleming
Washington:
Maggie Bell-McKinnon
Brian Healy
Rhonda Hunter
Kathy Jacobson
Jennifer Peters
Tim Lichen
Melissa Maxfield
Christopher Messina
Kirby Schaufler
Elaine Snouwaert
Laurie Usher
Diane Westfahl
Julie Winchell
W. Virginia:
Toni Lynne DeVore
John Wirts
Wisconsin:
Sue McMillin
Debra McNabb
Deb McRae
Mary Pardee
Jackie Scharfenberg
Julie Speck
Suzanne Wade
Wyoming:
Tim Foley
Sue Perin
Elaine Ullery
Canada:
Norm Frost
Justin Toner
Paul Whitfield
Mexico:
Antonia Giancola
Guillermo Larios de Anda
Maria Teresa Leal-Ascencio
Alicia Lerdo de Tejada Brito
Pilar Saldana
Rita Vazquez del Mercado Arribas

A Tangled Web
Conducting Internet Research

Summary
Students practice using the Internet and evaluating Web sites in order to gather local information about water quality.

Objectives
Students will:
- develop strategies to access information on the Internet.
- evaluate Web sites using accepted criteria.
- research pertinent water quality topics.

Materials
- *Computers with access to the Internet*
- *Copies of **Web Check** Student Copy Page* (one per student pair)
- *Copies of **Healthy Water, Healthy People Profile** Student Copy Page* (one per student pair)

Background
Because there is a wealth of information about water quality on the Internet, users often need guidance to maintain focus in their research and to determine the credibility of sources. Web research requires not only finding the information but assessing its reliability as well. Because just about anyone can put just about anything on the Internet, student researchers must be vigilant when evaluating information gleaned from the Web. The adage "Buyer Beware," can be adapted to the

Grade Level:
6–12

Subject Areas:
Computer Science, Language Arts, Technology Education

Duration:
Preparation:
30 minutes

Activity:
one or two 50-minute periods; could be used as a homework assignment

Setting:
Classroom, Computer Lab, or Library

Skills:
Gather, Organize, Analyze, Interpret, Evaluate, and Present

Vocabulary:
Internet, watershed, evaluate, criteria

Internet as "Researcher Take Care."

One of the first clues to a site's credibility is the top-level domain of the site's URL (that is, **.com, .org, .net, .gov, .edu, .biz**), which gives a general idea of what kind of site it is. (Refer to the *Web Check Student Copy Page* for more information.) For example, the quality of **.edu** sites is variable. Usually university faculty posts helpful research articles and information, but this is not always the case. Another clue to watch for is a URL that carries a tilde (~), which denotes an unofficial personal page carrying no governmental or institutional affiliation.

In general, government, educational institutions, museums, reputable non-governmental organizations, or well-known writers' and researchers' sites are considered mostly reliable when they meet the criteria from the *Web Check Student Copy Page*. For instance, contributors researching *Healthy Water, Healthy People* found information from the United States Environmental Protection Agency, the U.S. Department of Agriculture Natural Resources Conservation Service, the Centers for Disease Control, and university extension sites very helpful and reliable.

When in doubt about the accuracy or objectivity of Internet material, it is best to search for additional corroborative information. Before using any information, it is a good policy to find at least two independent sources to confirm it.

The skills used in this activity help

students critically review information gathered from Web sites. This activity leads to a higher-level Web site activity called "Multiple Perspectives" (p. 55), which challenges students to conduct a WebQuest and to cooperatively approach a water quality issue from differing perspectives.

Procedure

Warm Up

Ask students why they use the Internet and for what purposes (music, games, news)? Tell them that in this activity they will research the Internet to answer questions about their local water quality.

Divide students into cooperative groups and ask them to list the advantages and disadvantages of using the Internet for research. For example, the ability to view a multimedia presentation of a topic may be an advantage. A disadvantage may be that some multimedia applications require software that the user doesn't have and can't readily obtain.

Have students share their lists with the rest of the class. After the presentations, ask students: Who is responsible for assessing the reliability of sites being used for research? (Remind students that the researcher is responsible for assessing the quality of Web sites and that each individual must determine whether or not to use questionable sites.)

Activity

Part I

1. Divide students into pairs and distribute the *Web Check*

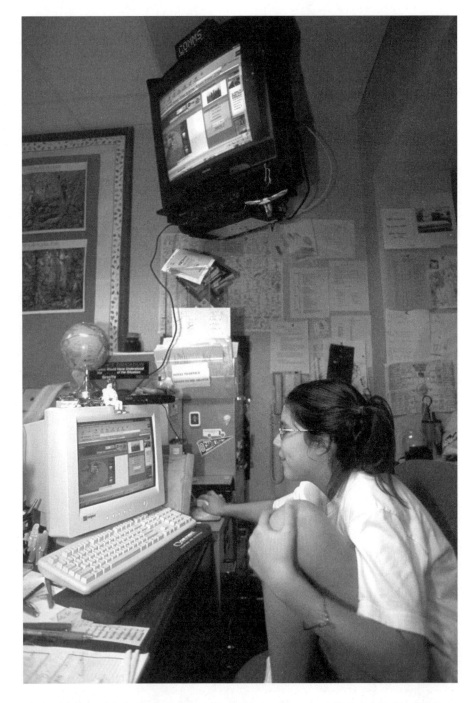

A student conducting research on the Internet. Courtesy: Richard Fields, Outdoor Indiana *Magazine*

Student Copy Page to each group and review each question with the class.

2. Ask them to contribute

additional *Web Check* questions that they feel are important when evaluating information from Web sites.

3. **Remind students that they will use these criteria to evaluate the Web sites used to answer questions in** *Part II.*

Part II

1. **Instruct students that they will use the Internet to answer questions about water quality in their local area.** If they are unfamiliar with the use of Internet search engines, review this information with them.

2. **Distribute the Healthy Water, Healthy People Profile and review the questions with students.** Ask students to use the Internet to answer the questions.

3. **Have students record their answers and the Web sites from which they found the answers.** Remind the students to use their *Web Check* questions as they conduct their research, and to record any Web sites that they disqualified because of the *Web Check.*

Upper Level Option:

To practice evaluating Web sites, instruct students to visit the *University of Idaho Library's Web Sites to Evaluate* Web page at http://www.lib.uidaho.edu/instruction/webeval.html. Have them choose a pair of sites from the list, write a report on the differences between them, and defend the one that they feel is most reliable.

Wrap Up

Review the students' answers to the *Healthy Water, Healthy People Profile.* How difficult was it to find the needed information? Was there more than one source of information for any of the questions? What Web sites were disqualified using the *Web Check* questions? Why?

Discuss with students the benefits and costs of using the Internet as a source for information about water quality. Have them review the *Web Check* again and ask them to contribute additional "checks." Ask students how their view of information found on the Internet changed as a result of this activity.

Assessment

Have students:

- develop strategies to access information on the Internet *(Warm Up, Part II).*
- evaluate Web sites for accurate and objective information *(Part I, Part II).*
- research pertinent water quality topics *(Part II).*

Extensions

Develop a more extensive set of water quality questions for the students to answer through Internet research, or challenge them to come up with their own questions. Have students research these questions and present their findings to the class.

Help students design a Web site to either disseminate accurate and objective information about water

quality in their local area, or to fill an information gap that they may have identified during their Internet research.

Resources

Henderson, J. R. June 13, 2000. *ICYouSee: T is for Thinking* (The ICYouSee Guide to Critical Thinking About What You See on the Web). Retrieved May 22, 2001 from the Web site: http://www.ithaca.edu/library/Training/hott.html

Prorak, D. September 27, 2001. *University of Idaho Library: Web Site Evaluation Criteria.* Retrieved June 17, 2002 from the Web site: http://www.lib.uidaho.edu/instruction/webcriteria.html

Prorak, D. February 11, 2002. *University of Idaho Library: Web Sites to Evaluate.* Retrieved June 17, 2002 from the Web site:// www.lib.uidaho.edu/instruction/webeval.html

The Watercourse. 2001. *Discover a Watershed: The Rio Grande/Rio Bravo Reference and Activity Guide.* Bozeman, MT: The Watercourse.

**In evaluating the usefulness of Web sites,
a student researcher may want to consider the following questions:**

1. **Who owns the site?**
   - .edu: sponsored by educational institutions, especially colleges and universities. Typically carries acceptable and scholarly information. A tilde (~) in the address may signify a personal Web page.
   - .gov: owned by the U.S. government. Typically used to highlight information printed in brochures and pamphlets. Information found here is generally reliable.
   - .org: owned by an organization—not necessarily a nonprofit. Can be good sources of information, but may carry an agenda.
   - .mil: owned by a branch of the U.S. Military.
   - .com: commercial Web sites that are maintained by a company. May be reliable, but not necessarily scholarly.
   - .net: first used for administrative purposes, now available for anyone. Wide variety of owners, and not necessarily scholarly.
   - .biz: business Web sites that are available to anyone. May be reliable, but not necessarily scholarly.

2. **Who are the authors of the site and why did they write it?** What are the author's credentials? What qualifies the individual(s) to comment on this topic? Are they trying to provide information (perhaps the result of their research), promote their point of view, or sell something?

3. **What is the format of the information?** Is it a journal article, popular magazine, newspaper report? Academic journal articles are generally peer-reviewed and the author's research methods and conclusions assessed before publication.

4. **Does the author provide other references that help support his/her research or conclusions?**

5. **When was the site last revised?**

6. **Can the information from this site also be found at another site?** The ability to "get the same answer twice" via Internet searching is a good way to check information.

Other evaluation criteria that you have developed:

Healthy Water, Healthy People Profile

To develop your own *Healthy Water, Healthy People Profile,* conduct Internet research to complete the following tasks:

1. Go to the United States Environmental Protection Agency's "Surf Your Watershed" Web Site at http://www.epa.gov/surf. Locate your watershed using any of the several methods offered on this page (e.g., zip code, city, county, etc.). Find the name of your watershed and record it. Then click on the "Index of Watershed Indicators" from that page and record the score from the scale on the right of that page.

2. Find the drinking water quality report issued by each city's drinking water treatment facility. If your city's report is not available on the Web, look for one from larger cities nearby. Within the report, there will be an abbreviation of "MCL." Record the meaning of the abbreviation. If your city's report is not readily available through an Internet search, try visiting http://www.epa.gov/safewater/dwinfo.htm and click on your state.

3. Conduct a search to generate a listing of water quality issues in your state (e.g., search for "your state + water quality issues"). Choose an issue and record the different Web sites that are related to it. From a reliable source, summarize the issue in a short paragraph. Note if you found any "unreliable" sources of information related to this issue and why you feel it is unreliable.

Mapping It Out:
Concept Mapping for Water Quality

Summary
Students use two teaching techniques—the KWL Process and Concept Mapping—to discover what they know, what they want to know, and what they learned about water quality. This activity is an excellent prior knowledge probe and assessment tool that can be applied to all of the activities in the *Healthy Water, Healthy People Water Quality Educators Guide*.

Objectives
Students will:
- organize prior knowledge of water quality on a concept map.
- investigate what they want to know about water quality.
- participate in a *Healthy Water, Healthy People* activity (activities) to learn about water quality topics.
- assess the knowledge gained from a unit or lesson about water quality.
- compare and contrast their knowledge about water quality before and after their unit of study.

Materials
- *3 large sheets of poster paper* (large enough for students to see from anywhere in the classroom)
- *Black marking pen*
- *Red marking pen*

Background
Acquiring new knowledge is an interconnected, interdependent, and ordered

Grade Level: 6–12

Subject Areas:
All

Duration:
Preparation:
15 minutes

Activity:
Variable (one-four class periods)

Setting:
Classroom (possibly outside–depends on activity

Skills:
Gather, Organize, Analyze, Interpret, Apply, Evaluate, Present

Vocabulary:
concept map, assessment

process. Assessing students' prior knowledge or preconceptions is a vital component of effective teaching. Students' preconceptions may be accurate and can facilitate learning, or they may be inaccurate and can impede learning. The class's current knowledge of water quality will be depicted on a concept map. By developing this map before and after water quality lessons, students will gain a strong visual picture of the learning process. The two teaching techniques integrated in this lesson are:

1.   **KWL**–This process is based on three questions: (1) What do you **k**now? (2) What do you **w**ant to know? (3) What did you **l**earn? These questions will elicit information from students at the beginning, at the middle, and at the end of a unit of study. The ideas generated from each of the questions are listed and posted in the classroom for reference during the learning process.

2.   **Concept Mapping**–Concept maps give students and instructors a working diagram of the students' knowledge of a certain topic and show relationships between concepts. Ideas are generated by the students and arranged on large sheets of paper to give them a picture of the entire concept. Main concepts are represented with circles and the links between those concepts (often lines) are labeled to better illustrate their relationships (see illustration on p. 7). Concept maps often contain several main concepts that are denoted by large circles. These main concepts can interconnect in a network pattern (Lanzing, 1997).

Information documented on the concept map is used to plan activities and to make concepts more relevant to the learner. After studying a unit on water quality, the class will return to their previous concept maps to add new information and details with the red pen. The added color indicates all the new knowledge acquired and becomes a powerful visual reinforcement of the learning process.

Procedure

Warm Up

Post a large sheet of paper in front of the class. Inform students that they will participate in a simple exercise to demonstrate how Concept Maps are developed. Write the word "puddles" in a large circle on the sheet of paper (see illustration). Ask students, "How are puddles formed?" Students are likely to suggest that puddles are formed by precipitation. Write "precipitation" in a circle connected by a line to the "puddles" circle. Have students suggest a verb that describes the relationship of precipitation to puddles, and write that word on the line connecting the two concepts.

Ask students to list the forms precipitation takes as it falls. Write those forms (e.g., rain, snow, sleet, hail) in circles and connect them to the precipitation circle, again adding illustrative verbs along the connecting lines to demonstrate the relationship of precipitation to its forms. Ask students where these forms of precipitation originate. Write "clouds" in a circle connected to all of the forms. Continue this process as long as it takes for students to understand the process of concept mapping.

The Activity

1. **After the class understands the concept mapping process, begin the water quality concept map. For this initial session, use only the black marker. Date a sheet of paper and label it "What I know about water quality" in a circle in the middle as in the Warm Up.** Have students tell you *what they already know* about water quality. Add these items to the concept map. List only information that they know, which may be very little at this time. Leave space for

additional subtopics that will be added during the assessment part of this lesson.

2. **Ask students to tell you things that *they would like to know* about water quality.** List these items on a separate sheet of paper titled "What We Would Like to Know About Water Quality."

3. **Ask the class if they have any suggestions about *how they can learn about* water quality.** Can they suggest references to use for this study? List their contributions on another sheet titled "How We Can Learn About Water Quality," and add your own. Tell the group that you are going to use

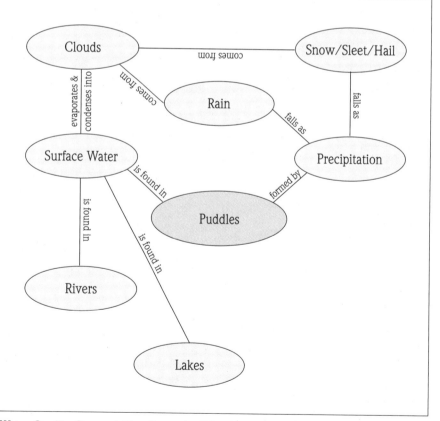

Water Quality Concept Map Example: What We Know About Water Quality

the information on both sheets of paper to plan a study unit about water quality.

4. **Develop a study unit about water quality using the concept map, your specific school curriculum, and the class lists generated in Steps 2 and 3.** The activities in this guide may address many of the topics and ideas that the students generated. *(Option: have students in small groups design their own water quality study units.)*

5. **When the unit has been completed, post the original concept map used in Step 1 of the activity.** Ask the class what they have learned about water quality. Using a red marker, date the paper again, and repeat the process of listing ideas. If you come across any incorrect information listed previously, cross out the error in red and write the correct information above it.

Upper Level Option

1. Instead of conducting The *Activity* as a class project, have small groups of students conduct the activity. They should proceed by following *Steps 1-5* as above. Offer guidance as needed. Instruct groups to develop a multimedia presentation of their materials and findings based on their answers to the questions in the *Wrap Up.*

Wrap Up

Examine the completed concept map and have students answer the following questions:

- What does the information

written in red represent?
- What did they learn about water quality?
- Did they learn everything they wanted to learn about this water quality topic?
- Can students estimate the percentage of knowledge gained (compare the amount of information in black and red)?
- What areas or topics had the most information?
- What topics had the least?
- Were there any inaccuracies in their initial knowledge about water quality?
- How did their ideas about water quality change?
- Why is it important to understand water quality topics?

Assessment

Have students:

- build a concept map (**Warm Up**).
- organize prior knowledge of water quality on a concept map (**Step 1**).
- investigate what they want to know about water quality (**Step 2**).
- participate in a *Healthy Water, Healthy People* activity (activities) to learn about chosen water quality topics (**Steps 3-4**).
- assess the knowledge gained from a unit or lesson about water quality (**Step 5**).
- compare and contrast their knowledge about water quality before and after their unit of study (**Wrap Up**).

Extensions

Have students investigate a local water quality issue that is related to the concepts they learned or wanted to learn in the activity. Possible ideas include creating a water quality education activity for their school by localizing the information from one of the Healthy Water, Healthy People activities; adapting and teaching a Healthy Water, Healthy People activity to students in younger grades; designing a water quality education campaign for their school or town; or becoming involved in a water quality monitoring or river cleanup project.

Resources

Angelo, T. A., and K. P. Cross. 1993. *Classroom Assessment Techniques.* San Francisco, CA.: Jossey-Bass Publishers.

Biehler, R. F., and J. Snowman. 1968. *Psychology Applied to Teaching. 5th Ed.* Boston, Mass.: Houghton Mifflin Company.

Gagne, E.D. 1985. *The Cognitive Psychology of School Learning.* Boston, Mass.: Little, Brown & Company.

Lanzing, J. 1997. *The Concept Mapping Homepage.* Retrieved June 28, 2002, from the Web site at http://users.edte.utwente.nl/lanzing/cm_home.htm

Novak, J. 1991. "Clarify with Concept Maps, a Tool for Students and Teachers Alike." *The Science Teacher* 58(7): 45-49.

WHAT IS
WATER QUALITY?

It's Clear to Me!

Summary

Students prepare various mixtures, investigate their unique properties, and then classify them using direct observation and the Tyndall effect.

Objectives

Students will:

- explore why some water pollutants are not easily seen using the Tyndall effect.
- prepare various mixtures in water.
- classify mixtures based on observed properties.
- suggest a method to distinguish salt water from tap water.

Materials

- *Beakers or glass jars* (five per group)
- *Water* (1 liter)
- *Milk* (.5 liters)
- *Paper* (heavy black or dark construction paper)
- *Scissors*
- *Slide projector, overhead projector, laser pointer, or flashlight*
- *Clear plastic cups*
- *Spoons* (1 per group)
- *Salt*
- *Glitter*
- *Cooking oil* (1 cup)
- *Masking tape*
- *Paper towels*
- *Copies of **Making Mixtures** Student Copy Page* (1 per student)

Grade Level:
6–12

Subject Areas:
Chemistry, General Science

Duration:
Preparation:
20 minutes

Activity:
50 minutes

Setting:
Classroom

Skills:
Observe, Record, Combine, Classify, Distinguish

Vocabulary:
Tyndall effect, emulsion, colloid, suspension, solution, solute, solvent, heterogeneous, homogeneous, mixture

Background

The relative quality of water is sometimes easy to ascertain. For example, when a glass of water smells sulphurous or when a lake is covered in dead fish it's a pretty sure bet something's not right. Other visible substances like sediments stay suspended in solution, giving a clue to the water's quality. However, it is often difficult to determine water quality using only our senses. Some pollution is invisible and odorless, like *Giardia* in a mountain stream, or lead in your drinking water. When a substance is dissolved completely in water, like sugar in tea, it becomes invisible. Understanding that many water pollutants are invisible is a critical step toward understanding the complexities of water quality. Below is a discussion of compounds, mixtures, and subsets of mixtures, all examples of classifications of water that contains substances or contaminants within it.

Matter is simply defined as anything that has mass and volume. Everything we see on Earth, even the air around us, is made of matter. All matter is made of atoms. Matter that is made of only one kind of atom and cannot be broken down into simpler substances is called an **element**. When two different kinds of atoms combine, they form a **compound**. Pure water, for instance, is a compound composed of oxygen and hydrogen.

Both elements and compounds are considered pure substances because they have a definite and uniform composition. For example, one atom of oxygen has the same composition as two atoms of oxygen, which has the same composition as three atoms of oxygen.

The smallest unit of a compound, the molecule, is made from a precise ratio of elements. A water molecule, whether in the form of ice, liquid, or gas is always composed of two hydrogen atoms and one oxygen atom.

Compounds containing other substances are called **mixtures**. While pure water (H_2O) is a compound, water found in nature is a mixture because it contains other substances dissolved in it. Water molecules with two hydrogen atoms and one oxygen atom are still present but dissolved gases (such as oxygen and carbon dioxide), and dissolved minerals (such as calcium and iron) are also present.

Mixtures can be described as either homogeneous or heterogeneous. Homogeneous mixtures are those that are uniform throughout, while heterogeneous mixtures are those that lack such uniformity. For instance, water is a homogeneous mixture known as a **solution**. Solutions contain a dissolving agent or solvent (pure water) and one or more dissolved substances or solutes (gases and minerals dissolved in the water). Because calcium is uniformly distributed in a glass of water, one sip will not contain more calcium than the next. Thus water is considered a homogenous mixture. Heterogeneous mixtures, on the

other hand, lack such uniform distribution of substances. Emulsions, suspensions and colloids, are examples of heterogeneous mixtures. An **emulsion** is a mixture of two liquids that will separate after standing. Oil mixed with water is an example of an emulsion. A **suspension** is a mixture of liquid and large, solid particles that can be separated by filtration or settling. Pepper in water or snow globes (round glass containers that are shaken to distribute glitter-representing snow) are examples of a suspension. A **colloid** is a mixture containing solid particles that are small enough to remain suspended, yet large enough to reflect light. Colloids may appear homogeneous to the unaided eye but under a microscope they prove to be heterogeneous. Milk is a good example of a colloid as the butterfat particles are small enough to remain suspended when left standing. In each case, components of heterogeneous mixtures are not uniformly distributed throughout. Separate spoonfuls of each would contain varying quantities of suspended particles (pepper or glitter in the suspension example, butterfat in the colloid example).

Below is an illustration of a partial classification of matter:

Procedure
Warm Up
Prior to the activity, fill two beakers or glass jars with two cups of tap water each. Add 2 drops of milk to one of them. Next, cut two one-centimeter diameter holes in a piece of dark paper so that one beaker can cover one of the holes and the other beaker can cover the other hole.

Begin class by posing this question to the students: Does the air in the classroom appear clean and clear? Can they see anything in the air that would lead them to believe it is not clean and clear? Most likely, they cannot.

Turn off the lights and close the blinds. Turn on a slide projector without a slide. (If you do not have access to a slide projector, use a laser pointer or a powerful flashlight.) The beam should allow students to see dust particles in the air. Explain that these particles are too small to be seen by the naked eye, but are large enough to reflect light. Tell students that this phenomenon is known as the Tyndall effect.

Turn on the lights. Put the paper with two holes in it on an overhead projector (or shine a laser pointer through the beakers). Place the beakers/glass jars on the paper over

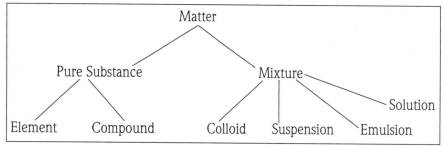

Illustration of the Tyndall Effect. Comstock Images.

the holes. Before turning the projector on, ask students if the beakers of water appear clean and clear. Can they see anything in the water that would lead them to believe the water is not clean and clear? Next, turn on the overhead projector. The light should shine through the holes in the paper and through the samples.

The sample without the milk should allow light to pass through it with no reflection. But similar to the dusty air example, suspended particles of milk in the other sample should reflect light and make it appear murky. This is another example of the Tyndall effect. Why does the sample with the milk appear murky? (Because milk is a colloid, a mixture containing particles [butterfat] small enough to remain suspended in liquid yet large enough to reflect light.) Can students think of other examples that demonstrate the Tyndall effect?

This activity will use the Tyndall effect and other means to distinguish and explore differences between various mixtures.

The Activity

1. Divide the class into small cooperative groups of about four students. Provide each group with five clear plastic cups half-full of water, a spoon, a teaspoon of salt, three drops of milk, a teaspoon of glitter, a tablespoon of cooking oil, five pieces of masking tape, a paper towel and a clear glass jar or beaker. Explain that students will use these materials to make various mixtures.
2. Ask students to use the pieces of masking tape to label their cups with the letters "a" through "e".
3. Distribute copies of *Making Mixtures* to each student. Have groups complete the worksheet.

Wrap Up

Discuss as a class the correct responses to worksheet questions.
- emulsion: oil and water
- colloid: milk and water
- suspension: glitter and water
- solution: salt and water
- solvent: water, solute: salt
- salt water and tap water are both solutions

Did the Tyndall effect help determine which mixture was a colloid? Could the Tyndall effect be used to distinguish between salt water and tap water? (No, salt is dissolved, therefore it does not reflect light). Write on the board the terms emulsion, colloid, suspension, and solution. Have students brainstorm additional examples of each type of mixture. Can they suggest other ways to distinguish between salt water and tap water?

Assessment

Have students:
- observe and then define the Tyndall effect (**Warm Up**).
- prepare various mixtures using water (**Step 3**).
- classify mixtures based on observed properties (**Step 3**).
- suggest a method to distinguish salt water from tap water (**Step 3**).

Extensions

Have students find examples of each type of mixture in their home. On a bulletin board post the items found for each type of mixture and have students check for accuracy.

Boil a second set of salt water and tap water to compare the boiling points (salt water boils at a higher temperature) or measure conductivity (charged salt and chlorine ions in salt water will conduct more electricity).

Testing Kit Extensions
Using the Healthy Water, Healthy People Testing Kits, measure the conductivity of the salt and tap water solutions. Compare the results. Also measure the boiling points of the salt and tap water solutions and compare results.

Resources

The American Chemical Society. 2002. *ChemCom: Chemistry in the Community.* New York, NY: W.H. Freeman and Company.

Simons, B., and T. Wellnitz. 2000. *Earth's Waters.* Upper Saddle River, NJ: Prentice-Hall, Inc.

Making Mixtures

1. You should have five cups, each half-full of water and each labeled individually "a" through "e". Prepare the following mixtures in these cups, drying the spoon with the paper towel between each mixture:

    a) Add 1 teaspoon of salt to the cup labeled "a" and stir until the salt dissolves. (Don't forget to dry the spoon after stirring!)

    Immediately after stirring, record any observations of your mixture in this cup:

    Five minutes after stirring, record any observations of your mixture in this cup:

    b) Add 2 drops of milk to the cup labeled "b" and stir vigorously.

    Immediately after stirring, record any observations of your mixture in this cup:

    Five minutes after stirring, record any observations of your mixture in this cup:

    c) Add 1 teaspoon of glitter to the cup labeled "c" and stir vigorously.

    Immediately after stirring, record any observations of your mixture in this cup:

    Five minutes after stirring, record any observations of your mixture in this cup:

    d) Add 1 tablespoon of oil to the cup labeled "d" and stir vigorously.

    Immediately after stirring, record any observations of your mixture in this cup:

    Five minutes after stirring, record any observations of your mixture in this cup:

    e) Add NOTHING to the cup labeled "e".

2. Read the definitions below and use them to fill in the chart on the following page. You may also wish to use the Tyndall effect to determine which mixture is a colloid. To do this, pour the mixture you want to check into a glass jar and place it over one of the small holes in the paper on the overhead projector (or laser pointer/flashlight).

Making Mixtures (continued)

Solvent: dissolving agent in a solution, usually the component present in the larger quantity.

Solute: dissolved substance in a solution, usually the component present in the smaller quantity.
Example: When sugar is dissolved in tea, the tea is the solvent and the sugar is the solute.

Homogeneous mixture: a mixture that is uniform or the same throughout.
Example: Sports drink from a carton. Each glass poured is the same as the last.

Heterogeneous mixture: a mixture that is not uniform throughout.
Example: Orange juice with lots of pulp, especially when not shaken. The first few glasses poured would contain less pulp than the last few glasses.

An **emulsion** is a mixture of two liquids that will separate after standing. Which of the above mixtures is an emulsion?

A **colloid** is a mixture containing solid particles that are small enough to remain suspended yet large enough to reflect light. Which of the above mixtures is a colloid?

A **suspension** is a mixture containing large, dispersed solid particles that can settle out and be separated by filtration. Which of the above mixtures is a suspension?

A **solution** is a homogeneous mixture that forms when one substance dissolves another. Which of the above mixtures is a solution? In the solution, what is the solute and what is the solvent? Under each letter, use the terms above to describe the type of mixture and whether each is homogenous or heteroenous.

	A	B	C	D	E
Type of Mixture					
Heterogeneous or Homogeneous					

3. **Of the above mixtures, two are actually solutions. Since gases (such as oxygen) and minerals (such as calcium) are naturally dissolved in water, even tap water is a solution. Set all samples but the two solutions aside. One of the group members must shuffle the two solutions while the rest of the group covers their eyes.**

Since both solutions are clear, how can they be distinguished from one another? Remember, NEVER TASTE CHEMICAL SOLUTIONS! Without tasting the two solutions, brainstorm ways to determine which is salt water and which is tap water.

From H to OH!

Summary
Students simulate the creation of acids and bases; manipulate acidic and basic solutions; and identify common acids and bases found at home.

Objectives
Students will:
- simulate a neutral solution, an acid solution, and a base solution.
- model the relationship between hydrogen and hydroxide ions to determine what makes an acid and what makes a base.
- create acid and base solutions by adding common household items to cabbage water.
- interpret the color change of a pH indicator solution to determine whether the solution is an acid or a base.

Materials
- *Colored paper* (2 different colors; 10 sheets of each color)
- *Purple cabbage* (medium-sized head) *and access to a stove or hot plate*
- *Small beakers, clear jars, or glasses* (4 per group)
- *Distilled water* (2 liters per class)
- *Eye droppers* (1 per group)
- *Plastic spoons* (1 per group)
- **Data Sheet** *Student Copy Pages* (1 per group)
- *Carbonated beverage–cola, soft drinks* (15 drops [1 ml] per group;

Grade Level:
6–8

Subject Areas:
Chemistry, Environmental Science, Mathematics

Duration:
Preparation:
30 minutes

Activity:
40 minutes

Setting:
Classroom

Skills:
Interpret, Organize, Communicate, Rank

Vocabulary:
hydrogen ion (H$^+$), hydroxide ion (OH$^-$), acid, base, logarithm, concentration

do not open until time of experiment)
- *Markers or tape to label sample containers*
- *Lemon juice* (15 drops [1 ml] per group)
- *Ammonia* (15 drops [1 ml] per group)
- *Baking soda* (½ teaspoon [1.3 g] per group)

Background
Because humans and aquatic organisms are dependent on water with pH levels within a range near neutral, pH is a crucial water quality indicator. The pH test, one of the most common and easily performed water quality tests, measures the concentration of hydrogen ions, which then allows us to infer the strength of the acid or base.

A water molecule (H_2O) can be thought of as one hydrogen ion (H$^+$) and one hydroxide ion (OH$^-$). An acid is defined as any substance that, when dissolved in water, increases the concentration of the hydrogen ion (H$^+$) (Ebbing, 1990). Therefore, an acid is a solution that has more hydrogen ions than it has hydroxide ions. A base is defined as any substance that, when dissolved in water, increases the concentration of the hydroxide ion (OH$^-$) (Ebbing, 1990). Hence, a base is a solution that has more hydroxide ions than it has hydrogen ions. A neutral solution has an equal number of hydrogen and hydroxide ions.

Mathematically speaking, pH is the negative of the logarithm of the hydrogen ion concentration (Ebbing, 1990). Hydrogen ion concentrations are very small numbers, so scientists developed the pH scale to make reporting and interpreting these numbers easier. Because the pH scale is logarithmic, for every one unit of change on the pH scale, there is a ten-fold change in the acid or base content of a solution (Mitchell, 1997). The pH scale ranges from 0 to 14, with a pH of 7 as neutral. An acid solution has a pH of less than 7, while a base solution has a pH of greater than 7. For example, acid rain with a pH of about 5 is ten times more acidic than natural rain with a pH of about 6. Likewise, a solution with a pH of 10 is one hundred times more basic than a solution with a pH of 8.

pH is usually measured by using either an electrode, which gives a digital reading, or an acid-base indicator dye, which changes color as the pH increases or decreases. The final color is then compared to a small chart or table to determine the pH.

Generally, natural waters have a pH of between 5 and 9 and most aquatic organisms survive in waters within this range. With the exception of some bacteria and microbes, if pH goes higher or lower than this range, aquatic life is likely to perish.

Other water quality problems can stem from high or low pH levels. Water with low pH increases the solubility of nutrients like phosphates and nitrates. This makes these nutrients more readily available to aquatic plants and algae, which can promote harmful overgrowth called "algal blooms." As these blooms die, bacteria numbers increase in response to the greater food supply. They, in turn, consume more dissolved oxygen from the water, often stressing or killing fish and aquatic macroinvertebrates.

Water with low pH can also corrode pipes in drinking water distribution systems and release lead, cadmium, copper, zinc and solder into drinking water. Water treatment plants vigilantly monitor the pH of the water that they treat to safeguard against such corrosion. Also water with a pH below 5 is too acid and above 9 is too basic for human consumption.

The pH of Common Items

Substance	pH or pH range	
Hydrochloric acid	0.0	Acidic
Stomach acid	1.0–3.0	Acidic
Lemon juice	2.2–2.4	Acidic
Vinegar	2.4–3.4	Acidic
Cola	2.6	Acidic
Grapefruit	3.0–3.2	Acidic
Acid rain	4.0–5.5	Acidic
Natural rain	5.6–6.2	Acidic
Milk	6.3–6.7	Neutral
Pure deionized water	7.0	Neutral
Sea water	7.0–8.3	Neutral
Baking soda	8.4	Basic
Milk of magnesia	10.5	Basic
Household ammonia	11.9	Basic
Sodium hydroxide	13.0–14.0	Basic

Table compiled from Ebbing, 1990, and Tillery, 1996.

Procedure
Warm Up

1. Review the pH scale, which ranges from 0 to 14. A pH of 7 is neutral. Most natural waters

Snow on a New England ski slope. Spring snowmelt can release a pulse of acidity from accumulated acid precipitation. Courtesy: Stephen Delaney; US Environmental Protection Agency

typically have a pH range between 5 and 9. Any solution greater than pH 7 is basic, and any solution less than pH 7 is acidic.

2. Explain to students that a water molecule (H_2O) can be thought of as one hydrogen ion (H^+) and one hydroxide ion (OH^-). The pH is determined by the concentration of hydrogen ions (H^+) in a solution. An acid is defined as any substance that, when dissolved in water, increases the concentration of the hydrogen ions (H^+) (Ebbing, 1990). Therefore, an acid is a solution that has more hydrogen ions than it has hydroxide ions. A base is defined as any substance that, when dissolved in water, increases the concentration of the hydroxide ions (OH^-) (Ebbing, 1990). Hence, a base is a solution that has more hydroxide ions than it has hydrogen ions. A neutral solution has an equal number of hydrogen and hydroxide ions.

3. Ask the students to help demonstrate the concept of pH in the following activity.

4. Write a large H^+ on ten sheets of paper of one color, and a large OH^- on ten sheets of paper of another color.

5. Ask two students to come to the front of the room. Give one student a sheet of paper with an H^+ on it, and the other a sheet with an OH^-.

6. Ask the other students to identify these ions—hydrogen (H^+), and hydroxide (OH^-). Ask if the solution created by these two students (H^+ and OH^-) is an acid,

base, or neutral. (This is a neutral molecule since H^+ has a charge of $+1$ and OH^- has a charge of -1.)

7. Ask two more students to join the other two students at the front of the room. Give one a sheet with H^+ and one a sheet with OH^-. Now ask one of the OH^- students to step aside for a moment. Is the solution still neutral? (Since it has a lone OH^- it has a net charge of $+1$, which makes it an acid.)

8. What will make the solution neutral? (The OH^- student needs to return.)

9. Next, ask one of the H^+ students to step aside. Now is the solution neutral? (Since there is a single H^+, the net charge is -1, so it is basic.)

10. What will make the OH^- solution neutral? (Bring the H^+ student back.)

11. Continue to experiment with this model, adding and subtracting more hydrogen and hydroxide ions, until you feel your students are comfortable identifying acid and base solutions.

12. Ask students what would be affected if streams or lakes had a pH

Mine acid drainage flowing from an open mine shaft. Courtesy: G. Donald Bain; The GeoImages Project; UC Berkeley

of less than 5 (e.g., fish and aquatic organisms would struggle to survive; land animals could not drink the water; etc.). Explain that, in the following activity, students will manipulate several solutions to determine whether they are acids or bases.

Upper Level or Mathematics Extension to Warm Up

To demonstrate the mathematical concepts of the logarithmic relationships of the pH scale, conduct a final step (most suitable for upper-level or mathematics students). Start with two students in front of the room—one holding an H+ card and one holding an OH- card. Ask the class to identify whether this solution is an acid, base, or neutral (the answer is neutral). Ask ten students to join them in the front of the room, all holding H+ cards. Again ask the class whether the solution represented is an acid, base, or neutral (it is an acid). Then ask them what number on the pH scale is represented by this solution. If they need a hint, let them know that the original two students represented a neutral solution, or 7 on the pH scale. A change in one unit up or down the scale requires the solution to be ten times more acidic or basic than the solution preceding it. The students are representing a solution that is ten times more acidic than neutral (10 more H+ ions), so the resulting unit on the pH scale is 6 (one unit more acidic than neutral 7). For bonus points, ask students how many H+

Teachers having fun representing hydroxide ions in the activity "From H to OH!" Courtesy: Project WET USA.

students are required to change the solution to a pH of 5 (10 x 10 or 100). For a pH of 4, 10 x 100 or 1000 H+ ions are required, and so on.

The Activity

You will be using cabbage water because it changes color as the pH changes, making it an acid-base indicator. Neutral cabbage water is purple, while acidic cabbage water is reddish and basic cabbage water is blue/green.

Note: It is important to use freshly made cabbage water as it quickly becomes unstable as an indicator. If possible, make the cabbage water on the same day you will use it, and refrigerate it until use. Also, it may be helpful to make a reference set of standards with color changes representing various pH ranges (e.g., 2, 4, 6, 8, 10, 12). The students can then compare their

color changes with these standards to determine approximately where on the pH scale their solutions reside. Another option is to measure and record the pH readings for the solutions so students can use them for later comparison.

1. **First prepare a cabbage water solution by boiling one medium-sized head of purple cabbage in 2 liters (more for larger classes or multiple classes) of distilled water until the water turns a deep purple color—approximately 20-30 minutes.** Filter or strain the cabbage and refrigerate the purple liquid.

2. **Divide the class into cooperative groups of three or four students.** Give each group four clear containers (each containing 25 milliliters of cabbage water), eyedroppers, spoons, and one *Data Sheet.*

© The Watercourse, 2003

3. **Ask the groups to add fifteen drops (1 ml) cola to one of the cabbage water jars, and stir. Record the color change and conclusions on the *Data Sheet.*** Label this sample #1.

4. **Take another 25 ml sample of cabbage water; add fifteen drops (1 ml) of lemon juice and stir. Record the color change and conclusions on the *Data Sheet.*** Label this sample #2.

5. **Follow the same procedure, using fifteen drops (1 ml) of ammonia, and a ½ teaspoon (1.3 g) of baking soda, putting each in 25 ml samples of cabbage water.** Stir and record changes on the **Data Sheet.** Label them samples #3 and #4.

6. **Have each group arrange their samples in order from most acidic to most basic (lightest red to darkest blue/green).**

7. **Instruct students to complete Part II on the *Data Sheet.***

8. **Ask them to complete Part III on the *Data Sheet* before beginning the *Wrap Up.***

Wrap Up

Ask students why one solution is more acidic than another? (Refer back to the **Warm Up** exercise.)

What would happen to the acid solution if a base, such as ammonia, were added? (With the addition of OH, the solution will approach neutral, and eventually pass it and become a basic solution. You may wish to demonstrate this with the class.)

Likewise, a single OH makes a solution basic. What would happen if an acid (vinegar), were added to a base solution? The solution will approach, and eventually surpass, neutral (pH = 7). Students may wish to experiment with their acid and base solutions by adding a neutralizing solution to their samples to reach a neutral pH in each sample.

Have student groups list ways that pH is important in their lives (e.g., digestion of food in our stomachs; herbivore digestion of plant matter [cows, deer]; acid rain; etc.) Have each group present their list and form a master list on the board.

Assessment

Have students:

- simulate a neutral solution, an acid solution, and a base solution with H and OH (**Warm Up**).
- model the relationship between hydrogen and hydroxide ions to determine what makes an acid an acid and a base a base (**Warm Up**).
- create acid and base solutions by adding common household items to cabbage water (**The Activity**).
- relate the color change of a pH indicator solution to determine whether the solution is an acid or a base (**The Activity**).

Extensions

Make your own litmus paper by dipping paper towels or coffee filters in the cabbage water. Allow the paper to dry, then cut into strips. Use your homemade litmus paper as an acid/base indicator by dipping it into several household products.

Using litmus paper (available from any hot tub retailer), test the pH of various household products. Make a chart of your results by lining up the litmus paper by color or acid/base concentration.

Invite a scientist from your local water treatment plant to discuss how and why he or she measures pH in drinking water.

Resources

Ebbing, D. 1990. *General Chemistry, Third Edition.* Boston, MA: Houghton Mifflin Co.

Hill, M. Date Unknown. *Determining Acids and Bases.* http://www.col-ed.org/cur/sci/sci119.txt

Mitchell, M., and W. Stapp. 1997. *Field Manual for Water Quality Monitoring: An Environmental Education Program for Schools.* Dubuque, IA: Kendall/Hunt Publishing Co.

Tillery, B. 1996. *Physical Science, Third Edition.* Dubuque, IA: Wm. C. Brown Publishers.

Torgerson, J. Date Unkown. *OFCN's Academy Curricular Exchange-Science.* http://ofcn.org/cyber.serv/academy/ace/sci/cecsci/cecsci181.html.

Data Sheet

I. After adding a common household item to the cabbage water, stir briefly and record your findings in the following table.

Common Household Item	Color of Solution
Cabbage water, nothing added	
Sample #1 15 drops (1ml) of cola added	
Sample #2 15 drops (1ml) of lemon juice added	
Sample #3 A ½ teaspoon (1.3 g) of baking soda added	
Sample #4 15 drops (1 ml) of ammonia added	

II. Look at the color change of the samples in Part I and determine whether the color change indicates the solution is an acid, a base, or neutral. Discuss with your group where on the scale you think each sample belongs. For example, the pH of purple cabbage water with nothing added is 7. Write "purple cabbage water" under the neutral (7) on the scale. Write the sample number on the scale where you think each sample belongs according to their pH.

	Acid					**Neutral**				**Base**				
0	1	2	3	4	5	6	7	8	9	10	11	12	13	14

III. After completing the experiment, answer the following questions.

1. What makes the cabbage water change color upon adding lemon juice?

2. What makes the cabbage water change color upon adding ammonia?

3. What do you think would happen if you combined the lemon juice solution with the ammonia solution?

© The Watercourse, 2003

Pollution–Take It or Leave It!

Summary

With a roll of a die, students simulate the movement of water within the water cycle to explore the water cycle's role with respect to water quality.

Objectives

Students will:

- identify the various forms water takes as it moves through the water cycle.
- role-play how water moves through the water cycle.
- role-play beneficial and detrimental impacts of the water cycle on water quality.
- compare and contrast the effects of different water cycle locations on water quality.
- visually and verbally depict the movement of water.

Materials

- 9 *large pieces of paper* (or manila folders as they stand upright)
- *Beads, tokens, or other small items–pebbles, coins, etc.* (approximately 250)
- 9 *plastic containers* (cups or bowls)
- *Paper cup for each student*
- Copy of **Water Cycle Table** *Teacher Copy Page*
- 9 *boxes, about 6 inches (15 cm) on a side* (Boxes are used to make dice for the game—one box for each of the nine water cycle

Grade Level:
6-8

Subject Areas:
Environmental Science, Language Arts, General Science, Earth Science

Duration:

Preparation:
1.5 hour for one time dice construction

15 minutes for set up

Activity: 50 minutes

Setting:
Outdoors or Classroom

Skills:
Gather, Interpret, Mapping/Organize, Analyze, Interpret, Cause and Effect

Vocabulary:
aquifer, condensation, evaporation, hydrologic cycle, percolation, runoff, sublimation, transpiration

stations. Coffee mug gift boxes are ideal. Label the sides of the die as indicated on the *Water Cycle Table*. Adding more boxes at each station, especially at the cloud and ocean stations, can increase the pace of the game.)

- *Copies of **Water Journey Map** Student Copy Page* (2 per student)
- *9 copies of **Water Pollution Chart** Student Copy Page*
- *A whistle, buzzer, or other noise maker*

Background

The water cycle, also called the hydrologic cycle, is a many-faceted process by which water moves over, under, and upon the earth through processes like evaporation, respiration, transpiration, runoff, percolation, storage (glaciers, snow pack, aquifers, plant and animal systems), condensation, and so on. Driven by the sun, every stage in the process affects the earth and its inhabitants in some way.

When we think of the water cycle, it is easy to imagine a circle of water, first evaporating from the earth, then condensing as clouds, then precipitating as rain, snow or hail back to the earth. Though this is a good basic description of the water cycle, in reality it is much more complex and involves a myriad of physical forms and distinct stages, locations and processes.

Heat energy from the sun evaporates water molecules from oceans, rivers, lakes, soil, plants, and animals. As this water rises as vapor into the air, it cools and condenses. Eventually it precipitates as rain, snow or hail, and then falls back under the force of gravity to the surface of the earth. Once on the earth, water takes any number of paths. It can seep into the ground to be absorbed by the roots of plants or percolate further to become ground water. Water can stay on the surface as runoff to rush as rivulets, then streams, and finally as meandering rivers on their course to the sea. Along the way, animals (including humans) can quench their thirst with it, or it can be absorbed by plants, or diverted for other human use. If it falls as snow, it can be stored in that frozen state (snow pack, glaciers) until enough heat energy or evaporation releases it. Indeed these examples illustrate a principle concept about the water cycle: movement. Water, in its three forms of gas, liquid and solid, moves constantly through our world.

One of the most important effects of the water cycle is its impact on water quality. Depending on the stage and location, the water cycle can improve or degrade water quality. Though evaporation, percolation and freezing play a purifying role as impurities and contaminants are removed during these phases, other aspects of the cycle actually contribute to degradation of water quality.

While the water cycle doesn't actually create pollution, it can aid in the transportation and concentration of contaminants. Runoff has the greatest effect in this regard. During runoff, water flowing over the earth's surface can pick up sediments and dissolve natural contaminants like minerals and salts along with human-contributed contaminants like petroleum products, fertilizers, pesticides, sewage and more. Eventually these contaminants are transported to streams, rivers, aquifers and oceans.

Even precipitation can be an agent of water quality degradation. When condensed water in the clouds precipitates back to earth as rain, snow or hail, it can pick up contaminants in the air. It then carries them to the earth, and eventually to surface and ground water systems.

Surface water flowing over a waterfall is but one component of the complex hydrologic cycle. American Water Works Association

Acid rain is an example of this phenomenon.

The water cycle is a process much like a web linking the whole earth and all living species together. In combination with heat energy and gravity, it becomes an agent of tremendous and constant change.

Procedure
Warm Up

Give students a piece of paper and ask them to draw how water moves on the planet. Have them identify the different places water can go, and indicate the different processes of that movement (condensation, evaporation, transpiration). Have students share their illustrations. (Many students are likely to draw the "water circle": water rains from the clouds, falls onto the land, runs off into the ocean, and evaporates back into the clouds). Discuss with them the processes of condensation, evaporation, and transpiration.

The Activity
Part I

1. Tell students that they are going to represent water molecules moving through the water cycle.
2. Create nine stations: animals, clouds, glaciers, ground water, lakes, oceans, plants, rivers, and soil by writing the names on large pieces of paper or manila folders and placing

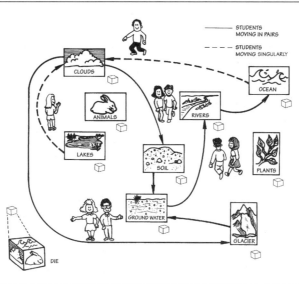

them at locations around the room or field. Place the previously constructed dice at their respective stations. (The *Water Cycle Table* provides an explanation of water movements.) Discuss the conditions that cause water to move. Explain that water movement depends upon solar energy and gravity.
3. Assign an even number of students to each station. Have students identify the different places water can go from their station by looking at the die at their station. Sometimes water will remain stationary. This is represented by the word "stay" on the dice.
4. Have students discuss the forms water takes when moving through the water cycle (i.e., solid, liquid, or gas). When they move as vapor they move alone to represent individual water molecules. When they move as liquid,

they will move in pairs to represent water molecules bonded together in a water drop.
5. In this game a roll of the die indicates where water goes. Students line up behind the die at each of the nine stations. At the cloud station they should be in single file; at all other stations they should line up in pairs. Students roll the die, and then go to the indicated location. If they roll "stay", they move to the back of the line. When students arrive at a new station, they go to the back of the line.
(Option: As students move to the cloud station, have them simulate the separation of water molecules during evaporation by separating from their partners and moving alone as individual water molecules. At the clouds station, have students roll the die individually. When they move as rain [condensation] out of the clouds, each student should grab a partner from the back of the line and move to the next location, again representing liquid. The partner does not roll the die.)
6. Students should keep track of their movements through the water cycle. Hand out copies of the *Water Journey Map* to all students and instruct them to draw lines to indicate their movement from station to station as they travel. If students roll "stay", they should record that

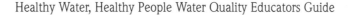

on their *Water Journey Map.* It is not uncommon for students to get "stuck" in one station for several turns. (*Option: Another approach is to place a container with different colored beads at each station to record their journeys. As students move from station to station, have them pick up beads and string them on lengths of twine or a pipe cleaner. For example, blue beads are offered at the ocean station, brown beads at the soil station, white beads at the glacier, and so forth.*)

7. Tell students the game will begin and end with the sound of a buzzer or whistle. Begin the game!

Part II

1. To demonstrate the role of water quality in the water cycle, have students conduct the activity again, this time to simulate the loss and gain of pollutants at each station. When they visit the different stations of the water cycle, students will collect and remove pollution "tokens" (beads, etc.) according to the *Water Pollution Chart.*

2. Place plastic containers and 50 tokens at the following stations: Soil, River, Lake, and Ocean. The tokens signify water pollution that the students gather as they move through the water cycle.

3. Place empty plastic containers at these stations: Animals, Ground Water, Plants, Clouds, and Glacier. These containers will hold pollutants (tokens) that are removed from water.

4. Assign a student to each of the nine stations to serve as the pollution monitor. These students are responsible for telling the water molecules when to collect or remove pollution (tokens) from their station depending on which station the student is moving from. Hand out copies of the *Water Pollution Chart* to help guide the station monitors. (*Option: Station monitors are not needed if large charts of the Water Pollution Chart are constructed for each station so the students can self-regulate their pollution collection and removal.*)

5. Hand out a Water Cycle Map and ask students to track their water cycle journey again. Give students a paper cup to hold their tokens as they collect and remove pollution on the water cycle journey.

6. Tell students that the game will begin and end with the sound of a bell, buzzer, etc. Begin play!

Wrap Up

Have students study their *Water Journey Maps* from ***Part I.*** Based on their journey, what can they conclude about the natural water cycle? Have them share their journeys with the class, using pertinent vocabulary words to describe their movements (e.g., "We *evaporated* from the ocean to the clouds, then *precipitated* into a lake, then *percolated* into the ground water, etc."). After running the simulation for ***Parts I and II***, ask them to illustrate or create a model of how they think

water moves on the planet. Have students compare and contrast their original and final impressions of the water cycle.

How many pollution tokens did they end up with in ***Part II?*** Did they collect more pollution tokens than they expected? What stations of the water cycle allowed them to remove pollution? Why? Which stations required them to collect pollution? Why? What are some examples of the pollutants that they collected? Based on this activity, can they suggest ways to improve water quality? (E.g., vegetative buffer strips along rivers and lakes could trap sediment before it reaches the water, carpooling or walking rather than driving can limit air pollution and acid rain, etc.)

Have students use their "travel records" to write creative stories about the places they visited as water molecules. Include a description of the conditions necessary for water to move to each location and what form water was in as it moved. Discuss any *cycling* that took place (if any students returned to the same station).

Assessment

Have students:
- identify the forms water takes as it moves through the water cycle (***Warm Up, Wrap Up***).
- role-play how water moves through the water cycle (***Part I and Part II***).
- role-play the beneficial and detrimental impacts of the

water cycle on water quality (*Part II*).

- compare and contrast the effects of different water cycle locations on water quality (*Wrap Up*).
- draw a picture or create a model to show how water moves on the planet (*Wrap Up*).
- write a story describing the movement of water (*Wrap Up*).

Extensions

Students can compare beneficial and detrimental impacts to water quality during different seasons and at different locations around the globe. They can adapt the game (change the faces of the die, add alternative stations, etc.) to represent these different conditions and locations.

Have students form groups of three —one atom of oxygen and two atoms of hydrogen (H2O). Ask one student to represent the oxygen atom by standing tall and still between the other two students. Have the other two "hydrogen" students act out their role as free-floating hydrogen atoms by running around wildly for fifteen seconds or so. When you say 'Bond!" the three students come together and lock arms, simulating a very stable molecule of water. Now, have them try the activity above with arms locked.

To simulate the effects of heat energy on water movement through the water cycle, introduce the following scenario several minutes into the activity. Have students stop where they are and ask them what happens to water when it is heated (molecules move faster). Have them proceed with the activity pretending the water in which they are traveling is now hot (but not boiling), meaning that they must move at a frenetic pace. Do the same with water that is almost frozen, where they move in super-slow motion. Then return to normal temperatures.

Use scenarios to illustrate the concept of water quality within the water cycle. While conducting *Part I*, have students stop where they are and read to them a scenario card that you have developed. For example: *A neighbor changed the oil in their truck and dumped it directly on their driveway. A rain came and washed the oil into the river, and it eventually polluted the lake that the river flows into. All students who are at the river or lake stations receive two pollution beads.* Develop other scenarios and pollution sources.

Have older students teach "*Pollution–Take It or Leave It!*" to younger students.

Resources

Alexander. 1989. *Water Cycle Teacher's Guide.* Hudson, N.H.: Delta Education, Inc.

The Watercourse. 1995. *Project WET Curriculum and Activity Guide.* Bozeman, MT: The Watercourse.

Notes:

Water Pollution Chart

Water Cycle Station	Water Quality Benefits and Impacts
If students move to...	Station Monitors instruct students to collect or remove pollution according to the following chart...
Soil	From Clouds–collect 1 for airborne contaminants From Animal–collect 2 for wastes
Plants	From Soil–remove 1 as some plants absorb pollutants from water.
River	From Soil–collect 2 for sediment from erosion From Lake–neither collect nor remove any tokens From Ground Water–remove 1 as ground water often is filtered by soil before it enters ground water aquifers
Clouds	From All Stations–remove 2 as water is purified as it evaporates
Ocean	From Clouds–collect 1 for airborne pollutants brought down with rain From River–collect 2 for pollutants carried in the river
Lake	From River–collect 2 for pollutants carried in the river From Clouds–collect 1 for airborne pollutants brought down with rain From Ground Water–remove 1 as ground water often is filtered by soil before it enters ground water aquifers
Animals	From Lake–neither collect or remove pollutants From River–neither collect or remove pollutants
Ground Water	From Soil–remove 1 as soil is a filter for pollutants From Lake–remove 1 as it moves through soil first, which filters pollutants From River–remove 1 as it moves through soil first, which filters pollutants From Glacier–remove 1 as it moves through soil first, which filters pollutants
Glacier	From All Stations–remove 2 as pollution is removed as water freezes

Water Cycle Table

STATION	DIE SIDE LABELS	EXPLANATION
Animal	two sides *soil* three sides *clouds* one side *stay*	Water is excreted through feces and urine. Water is respired or evaporated from the body. Water is incorporated into the body.
Clouds	one side *soil* one side *glacier* one side *lake* two sides *ocean* one side *stay*	Water condenses and falls on soil. Water condenses and falls as snow onto a glacier. Water condenses and falls into a lake. Water condenses and falls into the ocean. Water remains as a water droplet clinging to a dust particle.
Glacier	one side *ground water* one side *clouds* one side *river* three sides *stay*	Ice melts and water filters into the ground. Ice evaporates and water goes to the clouds (sublimation). Ice melts and water flows into a river. Ice stays frozen in the glacier.
Ground Water	one side *river* two sides *lake* three sides *stay*	Water filters into a river. Water filters into a lake. Water stays underground.
Lake	one side *ground water* one side *animal* one side *river* one side *clouds* two sides *stay*	Water is pulled by gravity; it filters into the soil. An animal drinks water. Water flows into a river. Heat energy is added to the water; water evaporates and goes to the clouds. Water remains within the lake or estuary.
Ocean	two sides *clouds* four sides *stay*	Heat energy is added to the water; water evaporates and goes to the clouds. Water remains in the ocean.
Plant	four sides *clouds* two sides *stay*	Water leaves the plant through the process of transpiration. Water is used by the plant and stays in the cells.
River	one side *lake* one side *ground water* one side *ocean* one side *animal* one side *clouds* one side *stay*	Water flows into a lake. Water is pulled by gravity; it filters into the soil. Water flows into the ocean. An animal drinks water. Heat energy is added to the water; water evaporates and goes to the clouds. Water remains in the current of the river.
Soil	one side *plant* one side *river* one side *ground water* two sides *clouds* one side *stay*	Plant roots absorb water. The soil is saturated, so water runs off into a river. Water is pulled by gravity, and filters into the soil. Heat energy is added to the water; water evaporates and goes to the clouds. Water remains on the surface (in a puddle or adhering to a soil particle).

Water Journey Map

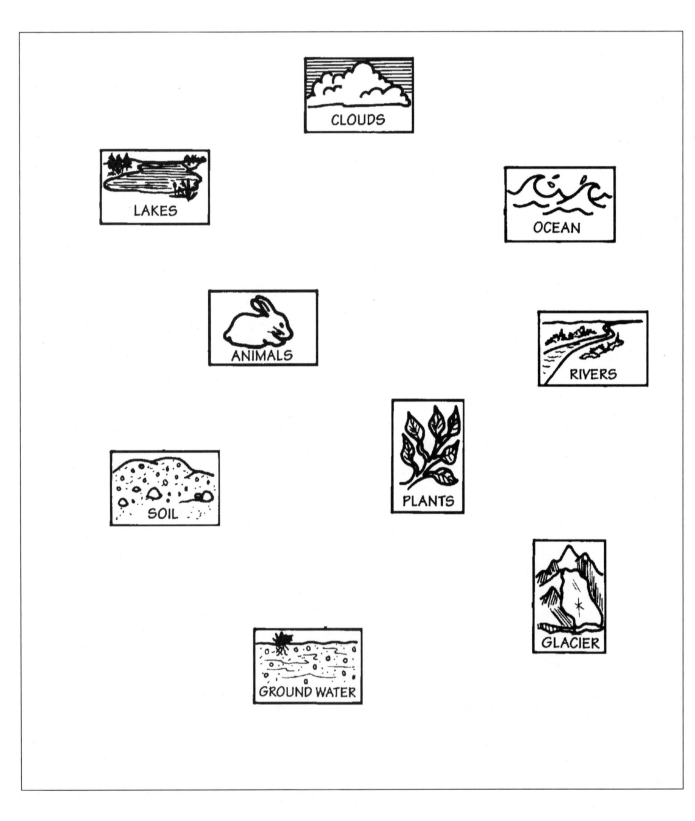

Grab a Gram

Summary
Students use familiar materials to gain an introduction to basic water quality measurements like parts per million (ppm) and milligrams per liter (mg/L). Students compare these measurements to national drinking water standards to determine toxicity levels of a contaminant.

Objectives
Students will:
- determine whether drinking water contaminants are always visible.
- use common objects to determine the relative size and weight of one gram, one milligram, and one microgram.
- calculate metric conversions from common water quality measurements.
- describe the relationship between common units of water quality measurement—mg/L, ppm, ppb.
- compare measurements of common water quality contaminants with allowable maximum contaminant levels of those contaminants.

Materials
- *Items that weigh about one gram* (sugar packet, raisin, paper clip, etc.)
- *Items that weigh more and less than one gram* (staple, clothespin, pen, etc.)
- *1 liter bottle filled with water* (1 per group) *(Optional: can be conducted as a demonstration with only 1 bottle needed)*
- *Approximately 1 kg (2.2 pounds) of sand, sugar, or salt (1 kg per group) (Optional: can be conducted as a demonstration with only 1 kg needed)*
- *Business card* (with at least 25 tiny pieces cut out of it) *(Optional: business card cut into 1000 equal pieces.)*
- *Copies of **Water Quality Measurement** Student Copy Page* (1 per student)
- *Copies of **Comparison of International Drinking Water Guidelines** Student Copy Page* (1 per student)

Grade Level:
6–8
Upper Level Option 9-10

Subject Areas:
Chemistry, Environmental Science, Earth Science, Life Science, Health Science, Social Studies, Mathematics, Metrics

Duration:
Preparation:
30 minutes

Activity:
50 minutes

Setting:
Classroom

Skills:
Organize, Calculate, Investigate, Analyze, Interpret, Measure, Compare and Contrast, Apply

Vocabulary:
concentration (of a solution), contaminant, drinking water standards, toxicity

Background
The effort to keep clean, healthy, drinkable water dates back to the Roman Empire when it was illegal to dump garbage in cisterns or to place a furnace, tannery, slaughterhouse, or cemetery within 25 meters of a well. Today, US cities and towns have drinking water treatment plants that monitor water quality daily. Every drinking water treatment plant produces an annual Consumer Confidence Report (CCR) that documents the average measured levels of each contaminant—chemicals, metals, or other substances that pollute water. CCRs are available to all consumers and are often printed in local newspapers and on the Internet.

Contaminants are measured to ensure that their *concentrations*—the mass of chemical per unit volume of water—do not exceed the Maximum Contamination Level (MCL) established by the US Environmental Protection Agency (USEPA). It is important to monitor these contaminants because in parts of the country certain contaminants have caused health problems when they exceeded the MCL. Concentrations of contaminants can vary widely, and some are simply far more harmful, or toxic, than others, even at very low concentrations (e.g., arsenic, lead, mercury, cyanide). Some contaminants that pose less risk to human health are called secondary MCLs. These include iron, fluoride, and aluminum. Though secondary MCLs are not federally regulated, many are regulated by state laws.

Most water quality measurements are recorded in milligrams per liter (mg/L), or the number of milligrams of a contaminant that are present in one liter of water. A milligram per liter (mg/L) is equal to a part per million (ppm) because a liter of

Dropping milligrams of a substance into a liter of water.

water weighs 1000 grams and a milligram is 1 one thousandth of a gram. Therefore, there are one million milligrams in a liter of water, and one of those milligrams equals one part per million.

Even smaller in concentration, a microgram per liter of water is 1 one thousandth of a milligram/L, and can be expressed as micrograms per liter, or mg/L. It is also equivalent to 1 part per billion (ppb). A part per billion is one thousand times a part per million.

Concentrations measured in ppm and ppb are so small that they can be difficult to conceptualize. Try these relative concentrations to help your students visualize them.
- One part per million is equal to one minute in two years.
- One part per billion is equal to one minute in 2000 years. (Water on the Web, 2001)

Conversions referred to in this activity:

1 kilogram (kg) = 1000 grams (g)
1 gram (g) = 1000 milligrams (mg)
1 kilogram (kg) = 1 million milligrams (mg)
1 kilogram (kg) = 1 billion micrograms (μg)
1 milligram (mg) = 1000 micrograms (μg)

Concentrations:

1 gram/L = 1 part per thousand (ppt)
1 milligram/L = 1 part per million (ppm)
1 microgram/L = 1 part per billion (ppb)

Procedure

Warm Up
Show students a clear glass of water.

Ask them if it is possible that there are chemical or biological contaminants in the water if they cannot see them? If students maintain that there may be contaminants in the water that they cannot see, ask them how to determine if there *are* contaminants in the water? Make a list of their answers.

After all answers are given, explain that water can be tested to determine if contaminants are present. These tests involve measurements of very small amounts of a contaminant. The most common unit of measurement in such a case is milligrams per liter (mg/L). Inform students that they will investigate common water quality measurements in this activity.

The Activity
Part I
(Note: Reforms in teaching the metric system call for students to use the metric system exclusively, without converting from the English system. This practice is applied in this activity. If conversions from English units to metric are needed, please see the Appendix for conversions or search the Internet for conversion calculators.)
1. Ask students to think about and predict the weight of several objects in the classroom. Have students record their predictions. If a scale is available, record the weight in metric units and compare them to their predictions. (Some metal folding chairs weigh about 4.5 kg. A new wooden pencil weighs about 7 grams.)

2. **Knowing that a pencil weighs about seven grams, ask students to name other class-room objects that weigh about one gram.**

3. **Pass around a tray of items, including some that weigh about the same as, more than, and less than one gram.** Have students try to select the items that weigh about one gram.

Part II

(Note: Part II can also be conducted as an instructor-led demonstration.)

1. **Divide students into small groups and distribute a copy of the *Water Quality Measurement Worksheet* to each student.**

2. **Provide each group with a 1 liter bottle and have them fill it with water.**

3. **Have students drop an**

Weight in Grams of Common Items	
Common Items	**Weight in Grams**
Thin rubber band	.16
Thick rubber band	.45
Small paper clip	.46
Plastic-head thumb tack	.47
Altoid™ candy	.79
Business Card	1.0
Large paper clip	1.2
Equal™ packet	1.2
Sweet and Low™ packet	1.3
Dime	2.3
Penny	3.0
Sugar packet	3.1
Wooden pencil-unsharpened	6.9

object weighing one gram (e.g., small paper clip, raisin, etc.) into their liter of water. Ask students what measurement or concentration results from this action? (1 gram in a liter of water, or 1 g/L). Have them record their answer on their worksheet.

4. **Have students examine their one-gram object, and imagine it sliced into 1000 pieces.** Prior to this step, cut 25 tiny pieces from a business card (business card weighs approximately 1 g), telling the students that you were "up all night painstakingly cutting a business card into 1000 equal pieces." Distribute a few tiny specks of paper to each group, informing them as you do, that each speck represents one thousandth of the original one-gram object.

5. **Ask students what measurement each of the tiny specks of paper represents?** (Answer: one thousandth of a gram, or one milligram [1 mg]). Have students record their answers on their worksheet.

6. **Have students drop one of the specks into the liter of water. What concentration or measurement results?** (Answer: one milligram per liter of water, or 1 mg/L). Ask students to record their answers on their worksheet.

Part III

(Note: Part III can also be conducted as an instructor-led demonstration.)

1. **Divide class into small groups and explain that 1 mg/L**

can also be expressed as one part-per-million. The following activity will demonstrate this concept.

2. **Explain the following conversions to the students. A milligram per liter (1mg/L) of water is equivalent to 1 ppm (part per million) because a liter of water weighs 1000 grams and a milligram is 1 one thousandth of a gram.** You may notice students struggling to understand these conversions, so explain that they will now be able demonstrate this concept for themselves.

3. **Give each group 1 kg of sand, sugar, or salt in a container or in a pile.** If you do not have a scale, use a little less than half of a five-pound bag (common size) of salt or sugar.

4. **Inform the students that the substance weighs 1 kg and for them to suppose that there are one million grains of the substance in the pile.**

5. **Instruct students to complete their worksheets for *Part III*.**

Upper Level Option

1. **Distribute the *Comparison of International Drinking Water Guidelines* chart to each group. Have each group complete the questions with the assumption that their liter of water contains 1 mg/L (1 ppm) of substance.**

Wrap Up

Review with the students the answers on the worksheets. (Answers: *Water Quality Measurement*

Worksheet, Part II: 1. 1g/L; 2. 1 mg; 3. 1 mg/L, *Part III*: 2. 1; 3. 1 ppm or 1 mg/L)

Were students able to successfully identify the correct measurements? Have students individually write out the list of measurements and conversions between concentrations (from kg to units u; from g/L to units u/L; from ppt to ppb).

Challenge students to create an analogy to describe one of the measurements from the activity (e.g., One part per million is equal to one minute in two years.) Share these analogies with the class.

Upper Level Wrap Up:

Have students discuss the differences in the allowable concentrations of arsenic, copper, nitrates, and sulfates in drinking water. What relationships exist between the concentration of a drinking water contaminant and the toxicity of the contaminant (risk to human health)?

Assessment

Have students:
- determine whether drinking water contaminants are always visible and apparent (**Warm Up**).
- use common objects to determine the relative size and weight of one gram, one milligram, and one microgram (**Part I, Part II, Part III**).
- calculate metric conversions related to common water quality measurements (**Part II, Part III**).
- describe the relationship

between common water quality measurement units–mg/L, ppm, ppb (**Part II, Part III**).
- compare measurements of common water quality contaminants with allowable maximum contaminant levels of those contaminants (**Upper Level Option**).

Extensions

Have students investigate their own city or town's drinking water quality by obtaining a local Consumer Confidence Report (CCR). Once obtained, compare the average local contaminant level with the Maximum Contaminant Level (MCL) for the contaminants studied in this activity.

Have students research the different measurements used for nitrates in the *Comparison of International Drinking Water Guidelines* chart. What is the difference between 10.0 mg/L as N and 50 mg/L as NO_3?

 Internet Extensions
Search for Web sites that contain Consumer Confidence Reports for your town or for the closest large city. Investigate any other pertinent information about measurement of water quality contaminant levels. Visit the USEPA Web sites for a complete list of MCLs in the U.S.

Have students visit *Water on the Web* at http://wow.nrri.umn.edu/wow/under/units.html Challenge them to find different ways of expressing ppm and ppb using ratios of common items.

 Testing Kit Extensions
Use Healthy Water, Healthy People Testing Kits to investigate the "detection limits" of water quality testing equipment. The lower the detection limit of the testing kit, the lower the concentrations that can be detected by the testing equipment.

Resources

Barzilay, J.I., et al. 1999. *The Water We Drink.* Piscataway, NJ: Rutgers University Press.

Hach Company. 1997. *Water Analysis Handbook.* Loveland, CO: Hach Company.

Harte, J., et al. 1991. *Toxics A to Z.* Berkeley, CA: University of California Press.

Water on the Web. 2001. http://wow.nrri.umn.edu

Instructors precisely titrating milliliters of reagent to determine the level of dissolved oxygen. Courtesy: Project WET USA

Water Quality Measurement Worksheet

Part I

1. Predict the weights of several objects found in the classroom:

Part II

1. Drop a one-gram object into your liter of water. What measurement or concentration results?

2. The instructor distributed tiny specks of paper and each was said to represent one thousandth of a gram. What is another way to refer to this measurement?

3. Drop one of the specks into your liter of water. What measurement or concentration results?

Part III

1. The 1 kg of substance before you contains one million grains.

2. Grab one grain of the substance. How many parts per million (ppm) of the substance do you hold?

3. Drop the grain into your liter of water. What concentration of the substance results from this? (Hint: You can express it two ways—as ppm or as mg/L.)

4. To convert parts per million (ppm) to parts per billion (ppb), grab another grain of substance. Imagine this grain of substance sliced into 1000 pieces.

5. Using a pen, touch the ink to a piece of paper to get the tiniest speck possible. Have a contest between the members of your group to see who can make the tiniest speck. Let this speck represent one microgram (even though a microgram is actually much smaller than this speck), which is 1000 times smaller than a milligram.

   Speck

   _____

6. One microgram per liter is equal to one part per billion (ppb). Some contaminants (e.g., arsenic) present health risks in such small amounts that they are often measured in parts per billion.

Comparison of International Drinking Water Guidelines

Refer to the *Comparison of International Drinking Water Guidelines* chart and answer the questions below:

1. Is 1 mg/L an acceptable level of:
   a. Arsenic in the U.S. according to the Environmental Protection Agency? _____
   b. Arsenic according to the World Health Organization (WHO)?_____
   c. Copper in the U.S. according to the Environmental Protection Agency?_____

2. Add grains of substance until you have a mixture of 11 mg/L. Is this an acceptable level of:
   a. Sulfate in Canada?_____
   b. Nitrates in U.S.?_____
   c. Nitrates in Canada?_____

Comparison of International Drinking Water Guidelines

Contaminant	USEPA	Canada	World Health Organization	Potential Health Effects	Sources of Contamination
Arsenic	0.05* mg/L (50 ppb)	0.025 mg/L (25 ppb)	0.01 mg/L (10 ppb)	Skin damage; circulatory system problems, risk of cancer	Erosion of natural deposits; runoff from glass and electronics production wastes
Copper	1.3 mg/L	1.0 mg/L	1·2 mg/L	Short term exposure: gastrointestinal distress. Long term exposure: Liver-kidney damage	Corrosion of household plumbing systems; erosion of natural deposits
Nitrates	10.0 mg/L as N	10.0 mg/L as N	50 mg/L as NO_3	"Blue Baby Syndrome": In infants under six months—life threatening without immediate medical attention	Runoff from fertilizer use: Leaching from septic tanks or sewage; erosion of natural deposits
Sulfate	250 mg/L	500 mg/L	250 mg/L	Gastrointestinal effects	Burning fossil fuels leads to atmospheric deposition

Adapted from: *Water Analysis Handbook.* Hach Company. Used with permission. * On February 22, 2002 the USEPA established the arsenic in drinking water rule, lowering the arsenic MCL to 10 ppb; the date by which drinking water treatment systems must comply with the new 10 ppb standard is January 23, 2006.

Stone Soup

Summary

Students will model and observe the acid neutralizing capacity of alkaline waters, and compare it with non-alkaline waters.

Objectives

Students will:

- model the acid neutralizing capacity of alkaline waters.
- recognize the characteristics of rocks that contribute to alkalinity.
- evaluate the effect of increased surface area on the reaction of acid on both granite and limestone rocks.

Materials

- *Assorted marbles* (one red, one blue, five white)
- *Clear plastic cup or glass jar*
- *Tape* (two different colors; e.g. red and blue)
- **Stone Soup Alkalinity Worksheet** *Student Copy Pages*
- *1 piece of limestone or concrete (or two antacid tablets) for each group*
- *Granite rocks for each group (can substitute glass or marbles)*
- *Two antacid tablets per group*
- *Vinegar (1 cup; or enough for 3 tablespoons for each group. Hydrochloric acid works if available— please consider safety precautions when handling hazardous chemicals such as this.)*
- *Tablespoon of quartz sand for each group*
- *Water*
- *Paper towels or paper plates*
- *Spoons or other items to crush antacid tablets*

Grade Level:
6–8
9–12 Upper Level Option

Subject Areas:
Chemistry, Life Science, Environmental Science

Duration:
Preparation:
30 minutes
50 minutes, Upper Level

Activity:
Option–50 minutes

Setting:
Classroom, Upper Level Option–Classroom or Laboratory

Skills:
Demonstrate, Model, Compare, Evaluate, Investigate

Vocabulary:
alkalinity, buffer, acidic, neutralize, ion, surface area

Background

Water constantly dissolves and absorbs the substances it contacts. This process begins as water passes through the air as rain, and continues when, as runoff, it contacts various minerals, soils and rocks. Rain absorbs carbon dioxide gas (CO_2) from the air, creating carbonic acid (H_2CO_3), a very weak acid. When this mild acid rain strikes the earth, it dissolves substances from the rocks and soils it contacts. The dissolution of minerals in these rocks can greatly affect water quality—especially pH levels.

The pH scale, which ranges from 0 to 14, is a measure of the concentration of hydrogen ions in a substance. The range between 0-6 indicates high concentrations of hydrogen ions, or an acidic pH range, while between 8-14 indicates low concentrations of hydrogen ions or an alkaline pH range. A pH of 7 indicates a neutral balance of hydrogen ions. Many fish and aquatic organisms are adapted to thrive in waters within a specific pH range, so any change in the acidity of that water can be harmful.

Such changes can occur as a result of acid rain (which is a by-product of burning fossil fuels [USGS, 2001]) or the introduction of certain industrial effluents. Natural substances can also shift the pH balance, but acid rain is of particular concern. Though car exhaust and industrial fossil fuel burning are mainly associated with urban areas, prevailing winds can carry acidified rain long distances into seemingly pristine areas. Whether natural water becomes more acidic due to acid precipitation depends on its location. Some waters will increase in acidity while others remain unaffected. How can this be?

The American Chemical Society reports that "researchers realized that bodies of water suffering from acidification due to acid rain have two features in common. First, they are downwind from a dense concentration of power stations, smelters, or large cities, which produce nitrogen oxides and sulfur oxides. Second, the bodies of water are often surrounded by soils that are unable to neutralize acid carried by the precipitation. If the soil cannot neutralize the acidic precipitation, the lake or stream into which the precipitation drains then becomes acidified. Bodies of water not seriously affected by acid rain often benefit from surrounding and underlying rock and soils that can neutralize the acidic precipitation. In particular, the effects of acidic precipitation can be greatly reduced by limestone ($CaCO_3$), which reacts with acids to produce soluble calcium bicarbonate" (American Chemical Society, 2002).

Limestone and other rocks that are composed primarily of calcium carbonate minerals are especially important to the balance of natural

water's pH. These materials when dissolved and joined with natural water bodies act as acid buffers and help protect these waters from the harmful effects of acid rain and other acid increasing influences. Water in these areas resist the effects of acidity, and species dependent on natural pH ranges are protected. On the other hand, no such buffering influence occurs in areas that lack calcium carbonate composition.

How does calcium carbonate affect this protective influence? When calcium carbonate ($CaCO_3$) is dissolved, ions of both calcium (Ca^{2+}) and carbonate (CO_3^{2-}) are released in the water. This reaction also produces carbon dioxide gas, which is released to the atmosphere (sometimes with a visible or audible fizzing). The chemical reaction of this process where calcium carbonate reacts with acids to produce soluble calcium bicarbonate is shown below: (American Chemical Society, 2002)

$$CaCO_3(s) + H_3O^+ \rightarrow Ca^{2+}(aq) + HCO_3^-(aq) + H_2O(l)$$

The positively charged hydrogen ions of any introduced acid are attracted to the negatively charged carbonate ions. These combine to reduce the overall concentration of hydrogen ions and thus acidity is neutralized. This neutralizing effect continues until the carbonate ions are used up. The capacity of water to neutralize hydrogen ions is called alkalinity (Murphy, 2000). This effect is increased as the surface area of exposed substances is increased thus allowing more calcium and carbonate ions to be released into affected waters. Water with a high

Helicopter drops lime in dying Lake Ovre to neutralize acid rain effects. Bergsjon, Sweden. Mark Edwards/Still Pictures/Peter Arnold, Inc.

© The Watercourse, 2003

buffering capacity, or alkalinity, is a safer place for fish and aquatic organisms than water without buffering capacity (KRAMP, 1997).

Of course some rocks lack this calcium carbonate composition and cannot act as an acid neutralizing influence. This helps explain why some lakes in predominately granitic areas are so susceptible to the harmful pH changes wrought by acid rain. Granites are composed primarily of the minerals quartz, feldspar, and mica which do not neutralize acid. Species adapted to specific pH ranges in these areas are particularly vulnerable to the affects of increased acidity.

Procedure

Warm Up

Note: This activity is designed to follow an activity illustrating the concepts relating to pH. If your students are unfamiliar with pH, please review the concepts of acid and base with them and/or conduct the Healthy Water, Healthy People activity From H to OH!

Part I

In a clear plastic cup or glass jar, place one red and one blue marble. Explain to students that the blue marble represents water, and the red marble represents hydrogen ions, or acidity. The acidic hydrogen ions (red marble) are always trying to react with the water (blue marble), causing the water to become more acidic. Gently shake the cup for fifteen seconds. Have students record on the **Stone Soup Alkalinity Worksheet** the number of times the red and blue marbles touch in those fifteen seconds.

Now add five white marbles to the cup. These represent alkalinity, or an acid buffer, which protects water from becoming acidic. Gently shake the cup again for fifteen seconds, with students recording the number of times the red and blue marbles touch. In which trial did the red and blue marbles touch the most? What

Mammoth Hot Springs in Yellowstone National Park. Limestone terraces are constantly changing in appearance. NPS Photograph. Courtesy: National Park Service

effect did the white, buffering marbles have on the water? In which trial was the water the least acidic?

The Activity
Part I
1. **Explain to the students that they will now demonstrate this same concept with an activity.** Remind students that this is a demonstration and not a competitive activity.
2. **Choose seven students: one student to represent Water, one to represent Acidity, and five to represent Alkalinity (or acid buffers). Have the remaining students form a tight circle approximately ten feet in diameter.** *(Option: placing desks or chairs in a circle also works, especially if you have a small group).* Hand one small piece of red tape to each of the students forming the circle, and instruct them to place it on their arm in such a way it can be grabbed from someone within the circle.
3. **Place both the Water student and Acidity student in the circle.**
4. **Explain that the Acidity student is always trying to reach the Water student, who is walking randomly from side to side in the circle, but not actively trying to avoid the Acidity student. The Acidity student is to grab a piece of tape, one at a time, from the students making the circle and place it on the Water student's arm.** This represents the water becoming more acidic.

5. **Have them do this for fifteen seconds, while the other students count the number of pieces of tape that the Acidity student manages to place on the arms of the Water student.** Record the results on the **Stone Soup Alkalinity Worksheet.**
6. **Repeat this activity, adding the five Alkalinity students into the circle. Place a piece of blue tape on the arms of the Alkalinity students.** The Alkalinity students are to randomly walk from side to side, but not necessarily next to the Water student.
7. **The Acidity student should again grab pieces of tape from the students that make up the circle. However, because acidity always reacts with alkalinity first when it is present, they must also grab a piece of tape from the Alkalinity student before** placing them both on the water.
8. **Conduct this activity for fifteen seconds.** Again, have students record results on the **Stone Soup Alkalinity Worksheet.**
9. **If time allows, repeat the activity, altering the number of acidity and alkalinity students.**

Wrap Up
Were there differences in the number of times the acidity could reach the water between the two trials? What caused these differences? Which of the trials would have the lowest (most acidic) pH and why? What role does alkalinity play in water?

Upper Level Option:
Part I
1. **Divide students into small cooperative groups. Provide each group with a dropper, three tablespoons of vinegar, a spoon, a piece of granite (or glass), a tablespoon of sand, and either a piece of limestone or concrete (or two antacid tablets).** Place the items on two paper towels or a paper plate.
2. **Explain to students that they will test the rocks for alkalinity.** If the rock is alkaline (contains carbonate), it will fizz when vinegar (or any acid) is dropped on it as carbon dioxide gas is released.
3. **Ask students to place two drops of vinegar on the piece of granite (glass) and record their results.** They should see no reaction, because the granite (glass) does not react with acid as limestone does.
4. **Ask students to place two drops of vinegar on the piece of limestone (concrete/whole antacid tablet) and record their results.** They should see the rock fizz because it contains carbonate ions, which react with acids to neutralize them and produce carbon dioxide gas and water.

Part II
1. **Ask students to experiment with how the surface area affects the reactions they just observed.** To accomplish this, they should place two drops of vinegar on the granitic sand (granite or quartz with a much greater surface area) (substi-

tute: marbles) and two drops of vinegar on a crushed-up antacid tablet (calcium carbonate or limestone with a much greater surface area). The fizzing should be even greater on the crushed calcium carbonate than on the whole calcium carbonate. No fizzing should occur on the granitic sand.

Upper Level Option *Wrap Up*

Did they observe a difference in the reactions between the acid (vinegar) dropped on granite and on limestone? Which rock acted as a buffer and which did not? When the increased surface area was exposed to the acid, did the previously observed reactions change? Why does limestone fizz when it is exposed to acid? What are the bubbles? [Carbon dioxide gas (CO_2).] This activity also works as a field investigation, with students determining whether a rock contains calcium carbonate by testing rocks in the field.

Assessment
Have students:
- demonstrate the buffering effect of carbonate in water using marbles and students to repre-

sent water, acid, and buffers (*Warm Up,* Part I: Steps 1–9).
- distinguish between a buffering rock and a non-buffering rock using vinegar (*Upper Level Option,* **Wrap Up**).
- compare the reactivity of granite and limestone to vinegar (acid) when surface area is increased (**Wrap Up**).

Extensions
Collect rainwater in a clear container. Put regular tap water in another container and some vinegar in a third. Place an antacid tablet in the rainwater, tap water, and vinegar to determine which is more acidic and to illustrate that rainwater is, indeed, slightly acidic. (Carbonate will react with acid by producing a fizz as carbon dioxide gas is released. Thus, if the antacid tablet fizzes acid is present.)

Explore the effect of temperature on dissolution by heating the vinegar and repeating the *Upper Level Option.*

Testing Kit Extensions
After soaking granite and limestone in beakers overnight, have

students test the water in each beaker for alkalinity using a Healthy Water, Healthy People Testing Kit.

Resources

American Chemical Society. 2002. *ChemCom: Chemistry in the Community, Fourth Edition.* New York, NY: W.H. Freeman and Company.

Kentucky River Assessment Monitoring Project (KRAMP). 1997. *Alkalinity and Water Quality.* Retrieved on November 29, 2001 from the Web site http://water.nr.state.ky.us/ww/ramp/rmalk.htm

Murphy, S. 2001. *General Information on Alkalinity.* Retrieved on December 3, 2001, from the BASIN Web site: http://bcn.boulder.co.us/basin/data/NUTRIENTS/info/Alk.html

Pipkin, B., and D. Trent. 1997. *Geology and the Environment, Second Edition.* Belmont, CA: Wadsworth Publishing Company.

United States Geological Survey (USGS). 2001. *Acid Rain: Do you need to start wearing a rainhat?* Retrieved on January 10, 2001, from the Web site http://ga.water.usgs.gov/edu/acidrain.html

Notes:

Part I: Marbles

First Trial:
In fifteen seconds, the red and blue marbles touched _____ times.

Second Trial:
In fifteen seconds, the red and blue marbles touched _____ times.

Wrap Up:
List possible reasons that the first and second trials were different:

What affect did the white, buffering marbles have on the water (blue marbles)?

In which trial was the water the least acidic?

Part II: Tape

First Trial:
In fifteen seconds, the number of pieces of red tape that the acidity student placed on the water is _____ pieces.

Second Trial:
In fifteen seconds, the number of pieces of red tape that the acidity student placed on the water is _____ pieces.

Wrap Up:
List possible reasons that the first and second trials were different:

What affect did the alkalinity students have on the water?

In which trial was the water the least acidic, or of lowest pH?

What role does alkalinity play in water?

PRINCIPLES OF WATER QUALITY ILLUSTRATE THE CONCEPT OF GOOD SCIENCE

Carts and Horses

Summary

The scientific method forms the foundation of good science, and science fair projects are conducted using this process. Inquiry-based learning is also related to the process of the scientific method. In this activity, students learn the scientific method in an inquiry-based activity, and then view examples of science fair projects that can be conducted with Healthy Water, Healthy People materials.

Objectives

Students will:
- investigate the steps of the scientific method.
- apply the scientific method to design an inquiry-based experiment.
- conduct a simple experiment demonstrating the steps of the scientific method.
- present their experimental design and results to the class.
- apply the concepts of inquiry and the scientific method to develop a science project (optional).

Materials

- Copies of **Designing Your Own Experiment** Student Copy page (1 per group)
- Copies of **Healthy Water, Healthy People** Science Fair Projects Student Copy Page (1 per student-Optional)

Grade Level:
6-12

Subject Areas:
General Science, All Science Subjects

Duration:
Preparation:
30 minutes

Activity:
two 50 minute periods

Setting:
Classroom

Skills:
Gather, Design Experiments, Analyze, Present,

- Glass or clear cup filled with tap water
- Clear plastic cups (1 per group)
- Coins (1 per group)
- Eyedropper (1 per group)

Background

Putting the cart before the horse is an old adage describing something that does not work because it is poorly planned. It is better to put the horse before the cart; in other words, plan a project before you start. Proper planning is important for most endeavors, but especially for investigations such as science projects and experiments. The scientific method is a planning tool for investigating the world around us in a meaningful way.

There are many observations of our world that we cannot explain and questions that we cannot answer. However, the scientific method gives us the tools to say "I don't know the answer but I can find out!" Questions about water quality are best answered using the scientific method. For example, if someone hands you a glass of water and asks if it is safe to drink, what do you say? How do you know whether it is safe to drink or not? Is it clear? If it is clear, does that also mean that it is clean?

In this example, you cannot assume that water is safe just because it is clear, because many contaminants, including bacteria and heavy metals, are either microscopic or dissolved in solution making them invisible to the

Steps of the Scientific Method:

Steps of the Scientific Method	Example
Observation *Identification of problem or question*	The two sides of a coin are different. Water drops can accumulate on the side of a coin without spilling. The tails side has smaller images.
Hypothesis *A prediction of the expected result*	The tails side of a coin will hold more drops of water than the heads.
Procedure *How you will test hypothesis*	Drop individual drops of water from a vertically-held eyedropper, one inch above the coin, onto both sides until they spill over. Repeat 3x per side. Count the drops and average each side.
Data Collection and Analysis *Take detailed notes of your observations*	Conduct the test according to the procedures. The tails side holds more drops.
Conclusions *Never **prove** a hypothesis—confirm or deny a hypothesis*	Confirm/accept the hypothesis as the tails side of the coin held more water drops than the heads side.

unaided eye. However, you can design an experiment to test the water to determine whether it is safe to drink. This process of designing an experiment to answer a question or investigate an observation is the essence of the scientific method.

The scientific method is the process by which all good science is conducted. It consists of a series of steps that start with a guiding question. A hypothesis, or prediction of expected results, is then developed. A procedure for testing the hypothesis is then designed and conducted. The

process ends with a conclusion about whether an experiment confirms or denies the hypothesis based on the data collected. The steps of the scientific method are outlined below, with a simple example of an experiment following the steps of the scientific method. This experiment will serve as an example of what the students will conduct during **The Activity**.

The scientific method is an important skill for all students to acquire as it relates directly with the work of scientists and researchers. The

process of "inquiry-based" learning mirrors the scientific method and is an excellent method for engaging students in the process of asking a question and learning how to answer the question using basic scientific principles.

Inquiry-based learning is "inquiry into authentic questions generated from student experiences and is the central strategy for teaching science" (National Research Council 2000). According to Martin-Hansen (2002), "Inquiry refers to the work scientists do when they study the natural world, proposing explanations that include evidence gathered from the world around them. The term also includes the activities of students—such as posing questions, planning investigations, and reviewing what is already known in light of experimental evidence—that mirror what scientists do."

Science fairs are another popular method for engaging students in the scientific method. Science projects that are designed to answer a guiding question or explain an observation duplicate the process that scientists go through when conducting experiments. The steps in developing a science fair project are similar to those in the scientific method, with a few additions (see sidebar). Please review the **Healthy Water, Healthy People Science Fair Projects** *Student Copy Page* for science project ideas.

There are several types of *inquiry* according to the National Research Council, (2000) including:

1. ***Open or Full Inquiry:***
Reflects what actual scientists are doing when they conduct research on a topic (scientific method). They start with a guiding question or unexplained observation, they develop a prediction or hypothesis about the question, they design a procedure to test their hypothesis, and they draw conclusions based on their results.

2. ***Guided Inquiry:***
Similar to Open Inquiry except that the guiding question or observation to be studied is chosen for the students, often by the teacher. The process of investigating the guiding question is then conducted by the students.

3. ***Coupled Inquiry:***
A combination of the Open and Guided Inquiry processes, beginning with a teacher-aided guided inquiry followed by a student-directed open inquiry.

4. ***Structured Inquiry (Directed Inquiry):***
Teacher-directed investigations or procedures which leave the experimental design process up to the teacher and not the students. The investigation becomes more of a "recipe" of procedures for the students to follow, disengaging them from the process of investigation design. This process is best used to introduce or model the process of inquiry by taking the students through an example investigation.

Procedure

Warm Up

Hold up a glass of tap water (but do not tell them the source) so the students can see that it is clear. Ask them if they think the water is safe to drink? How do they know that it is safe? Explain that there are many contaminants in water that are invisible to the naked eye, including pathogenic bacteria and harmful chemicals. Since these contaminants cannot be seen, ask again how can they determine if the water is safe for drinking? The answer is by designing an experiment to test the water.

Basic Steps of a Science Fair Project:

1. ***Select a Topic:*** Using research skills, select a topic that you are interested in and would like to know more about.

2. ***Research Your Selected Topic In-Depth/Bibliography:***
Try an Internet search engine or your local library to find information about your topic, other experiments that have been conducted, and background that may help you design your experiment. Create a comprehensive bibliography of your resources as you conduct your research.

3. **Use the Scientific Method to Design Your Experiment:**
   a. ***Hypothesis:***
   Make an educated guess about what you think your results will be. It must be testable using your procedures.
   b. ***Procedure:*** Plan and describe in detail how you will conduct your experiment that tests your hypothesis.
   c. ***Conduct the Experiment:***
   Carefully collect the needed data to draw conclusions about your hypothesis–may be graphs, charts, photographs, etc. Take detailed notes of what you observe.
   d. ***Analyze Data and Draw Conclusioms:***
   Write down your results of the experiment, making sure that you represent the results honestly. Much scientific value has come from experiments that "failed" to support their hypotheses. State whether you confirmed or denied your hypothesis.

4. ***Construct an Exhibit or Display:***
Develop a clear, clean presentation of the overall experiment (steps a–d above), including eye-catching graphs, charts, and photographs.

5. ***Rehearse Your Presentation:***
Judges will be asking you questions about your project, so prepare by practicing your general overview of the project and results and anticipating possible questions.

Review with the students the steps of the scientific method (from **Background**). Ask them to brainstorm ways to test the water to see if it is safe to drink (e.g., culture bacteria in a Petri dish; use a testing kit to determine if heavy metals are present; etc.). Inform the students that they will be using the scientific method to design their own experiment in this activity.

The Activity
Part I
1. Divide the students into small groups and distribute a cup of water, a coin, and an eyedropper to each group.
2. Instruct the groups to design a simple experiment using some or all of this basic equipment. Brainstorm possible experiment ideas with the class and write them on the board. For example: How many water drops will fit on the coin? How far over the edge of the coin can the water extend? Which side of the coin will hold the most water? Can the coin float on the water? Which causes a higher splash—dropping the coin oriented vertically or horizontally?
3. Have students formulate a hypothesis about their question. For example: The heads side of the coin will hold more drops of water than the tails side.
4. Have students write out the detailed procedures needed to test their hypothesis. Remind them that these procedures must be detailed enough to be duplicated by other researchers.

Hach Company/Healthy Water, Healthy People Water Quality Science Project Award Winner, Alexandra Antonioli of Butte, Montana, proudly displays her 2002 science project. Courtesy: Alexandra Antonioli and Linda DeMinck; Montana State Science Fair.

5. Have them conduct the tests following their detailed procedures, ensuring that they carry out at least three trials for each method.
6. Instruct them to collect, record, and analyze the data from their test.
7. Have the students draw conclusions about their hypothesis; did they confirm or deny their hypothesis?
8. Instruct students to develop and deliver a presentation of their experimental design, procedures, and results to the class.
9. Prior to the presentations, develop with the students a list of criteria for the presentations. For example: clearly uses all steps of the scientific method; uses visual aids with presentation; all group members participated equally in presentation; etc. Have students in the audience grade the presentations according to these criteria, representing science fair judges.

Part II
Optional: Designing a Science Fair Project
Building on the activity in Part I, encourage the students to develop their own science fair projects. The Healthy Water, Healthy People materials are ideally suited to help students successfully complete a meaningful water quality science project.
1. Distribute a copy of the *Healthy Water, Healthy People Science Fair Projects* to each student. Review the list of possible science projects with the students.
2. Brainstorm other possible project

ideas and write them on the board.

3. Review with the students the Basic Steps of a Science Project from the *Background.* Challenge them to plan, conduct, and present their science projects at local and national science fairs.

Wrap Up

After the presentations, discuss as a class the different experiments that were designed using such simple equipment. If appropriate, have students establish different awards (e.g., most creative experiment; best presentation; best adherence to the scientific method; etc.) and have them vote on the winners.

Explain to students that one of the most powerful aspects of conducting research is that it often leads to more questions about what is happening in the experiment. Challenge students to write down any questions that have arisen as a result of conducting their experiment. Ask them to briefly discuss further experiments that they would like to conduct to answer these questions.

Prior to the presentations, instruct the students in the audience to write down the Hypothesis, Procedures (summary), Analysis (summary), and Conclusions from each presentation as an assessment to ensure that they understand the steps of the scientific method.

Assessment

Have students:
- investigate the steps of the scientific method (***Warm Up***).
- apply the scientific method to design an inquiry-based experiment (***Part I* Steps 1-3**).
- conduct a simple experiment demonstrating the steps of the scientific method (***Part I*–Steps 4-7**).
- present their experimental design and results to the class (***Wrap Up***).
- apply the concepts of inquiry and the scientific method to develop a science project (***Part II***).

Extensions

If a science fair has not been started at your school, encourage students to initiate one. This involves planning the event and finding judges, awards, and space for the projects to be displayed. Have their science fair feed into regional, state, national, and international science fairs by connecting with the statewide science fair coordinator.

Encourage older students to mentor younger students to help them with their science fair projects. The older students could model a presentation so the younger students could see what a quality project and presentation looks like. Older students could serve as judges at a science fair for younger students.

Resources

Experimental Science Projects: An Introductory Level Guide. 2002. Retrieved on February 18, 2002 from the Web site: http://

www.isd77.k12.mn.us/resources/cf/SciProjIntro.html

Intel International Science and Engineering Fair. 2002. Retrieved April 4, 2003 from the Science Service Web site: http://www.sciserv.org/isef/

Martin-Hansen, L. 2002. Defining Inquiry. *Science Teacher.* 69 (2): 34-37.

National Research Council. 1996. *National Science Education Standards.* Washington, D.C.: National Academy Press.

National Research Council. 2000. *Inquiry and the National Science Education Standards.* Washington, D.C.: National Academy Press.

SciFair.org: The Ultimate Science Fair Resource. 2002. Retrieved on April 3, 2002 from the Society of Amateur Scientists Web site: http://www.scifair.org

Your Science Fair Project Resource Guide. 2002. Retrieved on April 3, 2002 from the Internet Public Library Web site: http://www.ipl.org/youth/projectguide/

Designing Your Own Experiment

Refer to the steps of the scientific method as you design your own experiment using some or all of the materials given to your group. An example of a simple experiment that follows the scientific method is given. Fill in the blank chart below to help you design and conduct your experiment.

Steps of the Scientific Method	Example
Observation *Identification of problem or question*	The two sides of a coin are different. Water drops can accumulate on the side of a coin without spilling. The tails side has smaller images.
Hypothesis *A prediction of the expected result*	The tails side of a coin will hold more drops of water than the heads.
Procedure *How you will test hypothesis*	Drop individual drops of water from a vertically-held eyedropper, one inch above the coin, onto both sides until they spill over. Repeat 3x per side. Count the drops and average each side.
Data Collection and Analysis *Take detailed notes of your observations*	Conduct the test according to the procedures. The tails side holds more drops.
Conclusions *Never **prove** a hypothesis—confirm or deny a hypothesis*	Confirm/accept the hypothesis as the tails side of the coin held more water drops than the heads side.

Complete this chart to help you design your own experiment:

Steps of the Scientific Method	Your Experiment
Observation Identification of problem or question	
Hypothesis A prediction of the expected result	
Procedure How you will test hypothesis	
Data Collection and Analysis Take detailed notes of your observations	
Conclusions Never **prove** a hypothesis—confirm or deny a hypothesis	

Healthy Water, Healthy People Science Fair Projects

The Healthy Water, Healthy People materials are ideal for use with science projects. Below is a sample of possible science project ideas using these materials. At the bottom of the list, fill in your own science project ideas.

General Topics:

- Why does the (dissolved oxygen/pH/nitrate/phosphate/flow/turbidity/conductivity/etc.) change from one part of my watershed to another?
- What are the differences in water quality between a local stream and pond/lake, and why?
- What is the pH of the rain or snow that falls near my school? Is it different depending on the location, weather pattern, or season?
- What do the macroinvertebrates found in a local river tell me about its water quality?
- Which type of water treatment/filter works the best for removing different contaminants?
- Is my well water/tap water safe to drink?
- How can I design a simple water treatment system for my home/cattle/fish tank?
- How does the geology of my area affect the water quality of the local streams and lakes? (alkalinity/pH/turbidity)
- What types of contaminants are found in runoff water from local parks/neighborhoods/ highways/other?
- What is the best method for keeping eroding sediments out of a stream?
- Which is the difference in pH/salinity/conductivity of soils from a wetland/pasture/ forest/playground/park/ garden/etc. and why? What affect does this have on plants?
- Why does the dissolved oxygen content differ between water bodies, seasonally, and at different times of the day? How does this affect aquatic organisms?
- Which nutrient (nitrate or phosphate) travels through the soil and which is held there? How does this affect the water quality of wells?
- How does the pH of soil differ between sites that are under pine trees and those that are not and why?
- How does sediment enter a waterway and what affect does it have on the temperature and dissolved oxygen of that water?
- Which type of soap is most effective at killing bacteria on your hands?
- What ecological and water quality changes occur when different restoration practices are conducted in a stream (vegetation planted on banks, logs sunken in stream, pollution removed, etc.)?

List Your Science Project Ideas Below:

Hitting the Mark

Summary
Students investigate the concepts of accuracy and precision in data collection, and learn the importance of writing detailed procedures.

Objectives
Students will:
- distinguish between accuracy and precision.
- investigate the relationship of accuracy and precision as it relates to water quality data collection.
- write clear procedures and recognize the limitations of those procedures.

Materials
- *Copies of **Accuracy and Precision Illustrations** Teacher Copy Page*
- *Clay (enough for each group to have at least a 3" x 1" x 1" piece)*
- *Copies of **Target** Student Copy Page*
- *Meter stick or tape measure*
- *Pencils and paper*
- *Colored pencils, markers, or crayons (at least 3 different colors for each group)*

Background
Data is a part of our lives in more ways than we may realize. We are bombarded daily with data from our televisions, radios, and computers. Our vehicles use data to determine if we

Grade Level:
6 – 12

Subject Areas:
General Science, Environmental Science, Chemistry, Mathematics, Biology

Duration:
Preparation:
15 minutes

Activity:
Warm Up and *Part I:*
50 minutes

Part II:
30 minutes

Setting:
Classroom

Skills:
Experiment design, Procedure design, Analyze, Interpret, Collect Data, Apply, Present

Vocabulary:
accuracy, precision, representativeness, completeness, comparability, replicate

are low on gas. Data drives how much chlorine is added to our drinking water during treatment. Even our watches give us data about the time of day.

The accuracy and precision of data is critical to whether we use it or not. We feel comfortable relying on data from a watch that is working correctly and has been set to the actual time to give us the correct time. However, if a watch is set one hour faster than the actual time we know that it is inaccurate, but it is still precise because it moves at a consistent rate around the dial. Of course, the most valuable watch is one that gives the correct time all the time, or consistently yields data that is both accurate and precise.

Likewise, the most valuable water quality data is that which is both accurate and precise. Because data collection is such an important component of water quality monitoring, the United States Environmental Protection Agency (EPA) has established Data Quality Objectives (DQOs). These objectives, which include *precision, accuracy, representativeness, completeness,* and *comparability,* are used to determine whether data is meaningful and valid (Mayio, 1996).

Precision is how consistent measurements are with each other rather than with the actual value. For example, if you were trying to hit the bull's eye of a dartboard, precision is demonstrated when the darts are clustered together, but not near the bull's eye. In water quality testing, an example of precision is when three pH tests yield the same result, but that result is not close to the actual pH of the water. *Replicate* samples (at least two) should be taken at each site to determine the precision of sampling.

Accuracy is a measure of how close results are to the actual value. Again, using the bull's eye analogy, accurate darts would all be fairly close to—but not necessarily in a tight cluster in—the center of the bull's eye. In water quality testing, accurate results could mean that three pH tests all measured very closely to the actual pH of the water. To assess accuracy, standards or reference samples should be analyzed along with field samples and then compared to each other.

Representativeness addresses the degree to which samples taken from a body of water represent the environmental conditions of that water. For example, a sample collected immediately downstream from a wastewater treatment plant may not be representative of the entire stream.

Completeness is a comparison of the minimum number of samples necessary for meaningful data analysis and the number of samples originally intended to be collected. When sampling over an extended period, there are various factors (e.g., weather, accessibility of the site, lack of resources, etc.) that can interfere with collection dates and time. Therefore, it is wise to collect more samples than are actually needed in order to obtain meaningful data.

Comparability refers to how well data can be compared with other data from the same project or data from another project. When testing water quality, comparability is maximized when standardized or accepted protocols are used for sampling, analyzing, and reporting data. Local, state and federal water management agencies publish such protocols.

The reliability of water quality data depends on its accuracy and precision, as well as the concepts of data representativeness, completeness, and comparability. Accuracy and precision tend to increase when more sophisticated technologies are used in measurement. In water quality monitoring, there are many levels of technology available to measure water quality. These technologies range from simply looking at a jar of water in the sunlight to spectrophotometers that read minute changes in shades of color.

Procedure

Warm Up

Discuss the definitions of accuracy and precision. Have students brainstorm ways that accuracy and precision are important in their lives. Distribute copies of **Accuracy and Precision Illustrations** or make a transparency and place it on an overhead projector. Have students determine which of the illustrations is 1) neither accurate nor precise; 2) accurate but not precise; 3) precise but not accurate; and 4) both accurate and precise.

Explain that they will now create their own demonstrations of accuracy and precision.

The Activity

Part I

1. **Divide students into small groups and distribute one piece of clay and a *Target* to each group.**
2. **Instruct each group to create five balls of equal size and shape from the clay.** Different groups may have different sized clay balls.
3. **Explain that their task is to drop the clay balls as close to the center of the target as possible from a height of no less than one meter (3.3 ft.).** The target should be placed on a flat, hard (non-carpeted) surface such as a floor, table, or desk.
4. **Have the students experiment with different ways of dropping the balls onto the target to maximize their precision and accuracy.**
5. **When the individual groups devise a method that they feel is accurate and precise, instruct them to record detailed procedures for this method.** Using their written procedures, have the students conduct three separate trials and record their results directly on the target using three different colored pens to signify each trial.
6. **Have each group present their procedures and results.** Were their results both accurate and precise? Discuss with the students

what could explain any variability of the results between the groups.

Part II

7. Have groups switch places with another group so they are using another group's procedures, targets, and clay balls.

8. At their new location, instruct each group to carefully read the new procedures and conduct three trials using the new procedures and materials, and then record their results.

9. Have each group present **their new procedures and results.** Were there any differences in the accuracy and precision of the results between **Part I and Part II?** Ask if there were any difficulties following another group's procedures. What were the primary causes of these difficulties? Ask the students how difficulty in following someone else's procedures could lead to inaccuracy in water quality data collection.

Wrap Up

Have students write a paragraph describing the difference between accuracy and precision. Ask students how difficult it was to be accurate and precise using their own procedures. How difficult was it to obtain accuracy and precision of data using another group's procedures? Why was there a difference? How important are clear and understandable procedures to collecting useful data? When collecting water quality data—for example, when testing water that comes out of the school water fountain—how important is it that the data collected is accurate and precise? What could happen if the data collected was inaccurate and harmful bacteria were found in the water?

Assessment

Have students:

*   discuss concepts of accuracy and precision (**Warm Up**).
*   design their own experiment to demonstrate accuracy and precision (**Part I**).
*   write detailed procedures outlining the steps of an experiment (**Part I**).
*   collect and record data for comparison with other students' results (**Part I, II**).
*   discuss the limitations of procedure writing (**Wrap Up**).

Extensions

Brainstorm ways students could improve their accuracy and precision in this activity (e.g., eye-hand coordination, distance from target,

TEAM 3
Procedures:

roll the clay
measure the three feet
take down the notes of where the balls dro
pped
Observe where they are dropped

black was scattered
blue was accurate
red was persistant

Sample procedures from an activity participant.

lighting, practice, or other conditions that affect the results). Conduct the activity again using the methods for improvement that the students came up with. Was there a change in the accuracy and precision of their results using the improvements?

Upper Level Extension

How could "technology" improve their results? Explain that technology is often considered electronic or computer-oriented. However, the spoon was probably considered high-tech when it was invented. Have students brainstorm other tools in the classroom that qualify as "technology."

Using the same groups and materials from the original activity, distribute instruments of increasingly sophisticated "technology" to the groups. For example, distribute the following (or use other equipment available in your classroom):
- a plastic spoon to the first group
- a plastic spoon and a cup to the second group
- a plastic spoon and a plumb line (3' string) to the third group
- a plastic spoon, a plumb line (3' string), and a 1' paper towel tube to the fourth group
- a plastic spoon, a plumb line (3' string) and a 3' long tube to the fifth group
- and so on as necessary

Have the students predict which groups will have the most accurate and precise results and then explain a rationale for their prediction. Have them conduct three trials and compare the results with their predictions.

 Testing Kit Extension
Using Healthy Water, Healthy People water quality testing kits, have students test a water sample (e.g., drinking fountain) for a parameter (pH) using the same testing procedure and equipment. Were there any differences in their results? Compile all of the results

and ask students whether they are accurate and/or precise. How do they know they are accurate? Discuss the definitions of comparability and representativeness. Ask students if their data demonstrates these qualities. What else do they need to know to be sure?

Resources

Dates, G. 1995. *River Monitoring Study Design Workbook.* Montpelier, VT: River Watch Network.

Hach Company. 1997. *Water Analysis Handbook.* Loveland, CO: Hach Company.

Mayio, A. 1997. *Volunteer Stream Monitoring: A Methods Manual.* Washington, D.C.: United States Environmental Protection Agency.

Merriam-Webster. 2002. *Merriam-Webster Online Collegiate Dictionary.* Retrieved from the Web site on April 23, 2002. http://www.m-w.com/

Notes:

Accuracy and Precision Illustrations

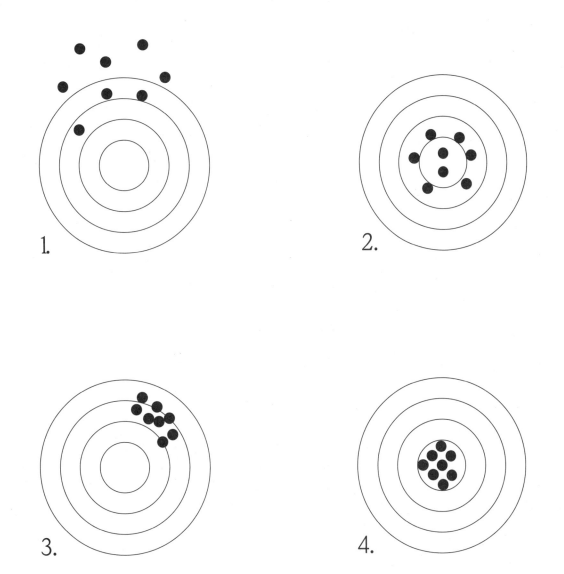

1.
2.
3.
4.

Adapted from *Water Analysis Handbook*. Hach Company: Loveland, CO. Used with permission.

Target

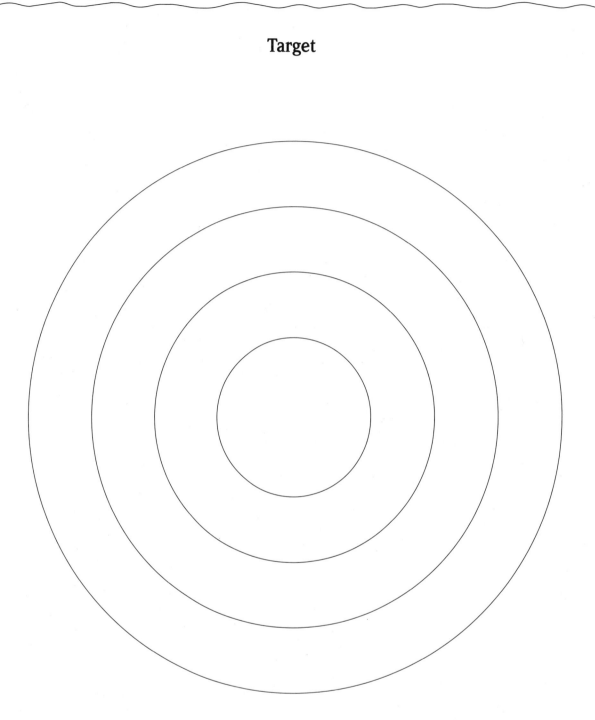

Multiple Perspectives

Summary

Students examine the perspectives of various water users, and participate in a WebQuest where they investigate a water quality issue and present solutions from different perspectives.

Objectives

Students will:

- interpret a data set from several perspectives.
- research a water quality issue using the WebQuest format of guided Internet research.
- present and defend a viewpoint using Internet and local resources.

Materials

- **WebQuest Student Worksheet** *Student Copy Page*
- **WebQuest Perspectives** *Student Copy Page*
- *scissors*

Background

Think of the many ways you have used water today. Perhaps you showered and brushed your teeth. You may have washed your breakfast dishes. Maybe you watered your lawn or garden? These are very basic, highly visible, domestic water uses.

But what about the uses you don't see or think about. For example, your electricity may be generated at a hydropower plant. Water is used to irrigate the fruits and vegetables you eat. What about recreation uses, such

Grade Level:
6–12

Subject Areas:
Social Studies, Language Arts, Mathematics, General Science

Duration:
Preparation:
10 minutes

Warm Up:
50-minutes

Activity:
two-three 50 minute periods

Setting:
Classroom

Skills:
Role-play, Associate, Interpret, Debate

Vocabulary:
maximum contaminant level (MCL),
part per million (ppm)

as swimming, water skiing, and fishing? Navigation and boating? Fish, animals, and microorganisms also need water to live.

Because there are many different uses for water, and each use demands different water quality standards, the United States Environmental Protection Agency (USEPA) created categories for specific water uses. These categories are:

- domestic and industrial use
- recreation for total body contact (swimming, water skiing, windsurfing)
- recreation for partial body contact (boating, fishing)
- aquatic organisms (fish, macroinvertebrates)
- agriculture (irrigation, livestock watering)
- commercial (navigation, hydroelectric and steam generated power)

Of course, depending on the water use, water quality data can be interpreted in as many ways as there are users. All water users—farmers, ranchers, industry, citizens, wildlife—have rights to water, yet all have different concerns about it. For example, farmers may prefer to irrigate with water containing high levels of nitrates, as it is a natural fertilizer. However, if water high in nitrates is used in infant milk formula, it can lead to methemoglobinema (blue baby syndrome), a condition that inhibits a baby's blood from carrying oxygen efficiently.

Because of this disparity in water quality interpretation, the USEPA and the Centers for Disease Control (CDC) have determined national standards for water quality which establish, for all water contaminants, a maximum contaminant level (MCL), or highest level of a contaminant allowed for each respective water use. Most of the contaminants are measured in parts per million (ppm), or milligrams per liter (mg/L).

Procedure

Warm Up

Ask students to brainstorm examples of data sets, which are simply a collection of data points. Examples include class grades, ages of the students in the class, daily low temperatures throughout the year, monthly precipitation levels for the year, etc. Using the example of monthly precipitation levels throughout the year, have students brainstorm different groups that are affected by the amount of monthly precipitation that falls. Examples could include farmers, construction workers, anglers, skiers, aquaculture (growing aquatic species for food, such as catfish), etc. Ask students to determine how monthly precipitation levels affect each group in their brainstorm, and if those groups each want the same amount of precipitation at the same time of year. Each group will typically have a unique perspective on the way they view data.

Have students demonstrate this principle of multiple perspectives of data using a picture as an example.
1. Use the photograph provided of President Roosevelt at Death Valley, California, or find a picture or photograph that is especially meaningful or diverse in its content. Examples include:
  a. Yalta Conference or other historical photograph of world leaders
  b. A scenic landscape picture containing a broad view of the land

President Franklin Roosevelt in Death Valley, California. Courtesy: National Park Service

2. Distribute or display the photograph and ask students to study the picture. Explain that there are many ways to interpret it, depending on your expertise and point of view. Brainstorm different perspectives for interpreting the picture and write the list on the board. The Death Valley photograph: historian, geologist, hydrologist, botanist, fashion designer, news reporter, etc.
Other possible examples include:
   a. Yalta Conference photograph: historian, artist, sociologist, architect, women's rights advocate, news reporter, etc.
   b. Scenic landscape picture: ornithologist, botanist, geologist, urban resident, developer, farmer, etc.
3. Break students into small cooperative groups and assign each a perspective from the list. Have groups consider the picture from their assigned perspective and write down how they view the image.
4. Have students present their views to the class. When all groups have presented, discuss their presentations. Is any interpretation more valid than another? Is one perspective right and another wrong? Was one group's perspective the same as another group's?

Explain to students that the photograph is a data set, and can be viewed from different perspectives.

Discuss how water quality has data sets that can, much like the photo, be viewed from different perspectives.

The Activity
1. Distribute one *WebQuest Perspectives Card* to each group.
2. Instruct the students that they will be participating in a WebQuest (Internet-based guided research) to find the information needed. Distribute the WebQuest Student Worksheet to each group and instruct students to complete it using the Internet resources provided. Encourage students to seek out other sources of information as well.
3. Conduct a town meeting where the Cooper Township Council hears presentations from all of the multiple perspectives on the issue of high turbidity in the local stream.

Wrap Up
Discuss the presentations as a class. Were any of the perspectives the same? How about the solutions? Were there common themes that surfaced from the different perspectives?

Ask students how the research process went. Were they able to find all the information they needed from

the Internet? Did they seek out local resources?

Assessment
Have students:
- interpret a data set from several different perspectives (**Warm Up**).
- research a water quality issue using the WebQuest format of guided internet research (**The Activity**).
- present and defend a viewpoint using Internet and local resources (**The Activity**).

Extensions
Include local issues in the town meeting by having one of the characters represent a local perspective. Ask a local official to participate in the role-play.

Resources
Hach Company. 1997. *Water Analysis Handbook.* Loveland, CO: Hach Company.

Mitchell, M. & W. Stapp. 1997. *Field Manual for Water Quality Monitoring: An Environmental Education Program for Schools.* Dubuque, IA: Kendall/Hunt Publishing Co.

Niles, R. 2001. *Data Analysis.* http://nilesonline.com/stats/dataanly.shtml

Notes:

WebQuest Perspectives

Polly/Paul E. Tishen, **City Manager** · http://wow.nrri.umn.edu/wow/under/ parameters/turbidity.html	Chip Silica, **Microchip Plant Owner** · www.hach.com/h2ou/h2wtrqual.htm#turbidity
Connie Sumer, **Consumer and Concerned Mother** · http://wow.nrri.umn.edu/wow/under/ parameters/turbidity.html · http://cru.cahe.wsu.edu/CEPublications/ eb0994/eb0994.html	Joan/John Dear, **Farmer and Nursery Owner** · www.verde.org/divert/wp319.html
	Dr. Samantha Gold, **Scientist** · www.fao.org/landandwater/watershed/ watershed/cases/brazil.pdf
E.P. Agenson, **Environmental Regulator** · www.epa.gov/safewater/mdbp/pdf/turbidity/ chap_09.pdf	Eddy Sanford, **Forester** · www.forestry.state.ar.us/bmp/roads.html · www.cnr.vidaho.edu/extforest/WQ6.pdf
Tom Driven, **Highway Department** · http://www.nmenv.state.nm.us/swqb/ cordovaTMDL.html	Mary Brown, **Treatment Plant** · www.dep.state.pa.us/dep/deputate/watermgt/ wsm/facts/fs2655.htm
George Fly, **Fisherman** · www.combat-fishing.com/streamecology.html	

WebQuest Student Worksheet

Using the Internet links provided, research the turbidity issue from the point of view on your WebQuest Perspectives Card by reading the background and completing the following questions:

Background:

Cooper Township is a small community (population 25,000) that is supported by several businesses, including construction firms for the growing community, farms, a small Microchip Production Plant, and local tourist services—fishing tackle shops, historical tour guides, craft shops, hotels, restaurants, and antique stores. Cooper Township has a large stream running through it and this stream supplies the township's drinking water. The watershed for the stream is entirely contained within the township, and the stream typically carries a heavy sediment load and is high in turbidity. The Cooper Township Council has called a town meeting to discuss the issue of high turbidity in the stream and the town's drinking water and what to do about it.

Your Task:

Research the issue of high turbidity in the local stream from the perspective on your WebQuest Perspectives Card using the Internet links provided. You may also use local resources, experts, and other resources as needed. At the end of the activity, your group will be asked to present your point of view of the high turbidity issue and to be able to defend your point of view in a public meeting. You will also be asked what your group recommends should be done to reduce the high turbidity levels in the stream.

Research Questions to Answer:

1. List whom you represent and any background information about your perspective from the Web Quest Perspectives Card.

2. Make a list of all the resources you used to research your perspective, including Internet sites, local resources, etc.

3. How strongly do you care about the high turbidity in the stream? Are you strongly opposed to it, moderately opposed to it, don't care, or support the high turbidity? Provide evidence to support your position.

4. What do you think should be done about the issue of high turbidity in the town's drinking water?

Presenting Your Point of View:

You will present your point of view in a town meeting in front of the mayor and city council members. Be prepared to give a professional presentation of your perspective (visual aids are recommended). Support your point of view with evidence you found in your research.

WATER QUALITY MUST BE MONITORED AND INTERPRETED

A Snapshot in Time

Summary
Students use a topographic map to explore the concept of a watershed and then apply that knowledge to watershed monitoring. Students will discern the differences in value between an individual data set collected at one place and time on a watershed versus a series of water quality data sets collected at various points along a watershed over time. Students will first graph watershed data then analyze, compare, and summarize trends in water quality.

Objectives
Students will:
- identify significant water quality parameters.
- read and interpret a contour map to determine the characteristics of a watershed.
- discern the value of data from a single sample as compared to data from a series of samples.
- graph, analyze, and summarize both spatial and chronological trends in water quality data.
- compare and contrast the effects of water quality parameters on one another, especially the relationships between temperature and dissolved oxygen, and the relationships between flow and clarity.

Materials
- *Pencils*
- *Graphing paper (or loose-leaf*

Grade Level:
6–9

Subject Areas:
Mathematics, Environmental Science, Language Arts, Geography, Chemistry

Duration:
Preparation:
15 minutes

Activity:
50 minutes

Setting:
Classroom

Skills:
Interpret, Analyze, Organize, Gather, Communicate, Present, Graph

Vocabulary:
headwaters, confluence, midriver, falls, downriver pH, nitrates, ammonia, phosphates, turbidity, dissolved oxygen, total dissolved solids, trends

paper if graphing paper is not available)
- ***Cooper River Watershed Map***
  Student Copy Page
- ***Watershed Data Summaries Worksheet***
- ***Cooper River Water Data Cards***
  Teacher Copy Pages

Background
Can you imagine living an entire lifetime without ever having a check-up with your doctor? Periodic check-ups can help doctors determine trends in our health and better avoid major health problems. The same is true of watersheds; they too need monitoring over time to assess trends in their health. Like a person who is generally healthy but has suffered a specific injury, say to the knee or shoulder, a river may be generally healthy except for a specific "injured" section such as a stretch contaminated by urban runoff, or one unduly susceptible to high sediment loading in spring. Watersheds need to be monitored in order to preserve their water quality health.

The United States Environmental Protection Agency (USEPA) cites these five major reasons for monitoring a river:
1. to "characterize waters and identify changes or trends in water quality over time;
2. to identify specific existing or emerging water quality problems;
3. to gather information to design specific pollution prevention or remediation programs;
4. to determine whether program

goals—such as compliance with pollution regulations or implementation of effective pollution control actions—are being met; and

5. to respond to emergencies, such as spills and floods."

Water quality is determined by conducting physical, chemical and biological measurements of a watershed. Chemical measurements include, but are not limited to, pH, dissolved oxygen, nitrates, phosphates, and dissolved solids. Some of the physical parameters often measured are clarity, temperature, discharge (flow rate), and the conditions of stream banks and shoreline. Biological assessments of water health measure the abundance and diversity of aquatic plants and animals.

The USEPA, along with many other local, state, national and international organizations, have found that if monitoring data is to be useful guidelines must be established and followed. One of the most important

Students investigating the water quality of a local stream. Courtesy: Richard Fields, Outdoor Indiana Magazine

guidelines states that strategically collecting numerous data samples that meet chosen objectives is a more effective approach than using data from a single sample, a practice that limits the findings that can be extracted from that data.

For example, it is most valuable to monitor a river several times throughout the year over a period of several years. In doing so, seasonal variations and annual variations can be better understood. This method also provides insight into the impact of pollution, development, and other land uses over time. It is also important that several sites along the length of the river be sampled so land use and environmental variations along the river can be documented.

Finally, the sampling parameters and the associated sampling techniques must be consistent over time and between different monitors. Samples should be collected and analyzed using standardized techniques that are well documented and accepted (USEPA, 1996). These measures ensure that data from one river can be readily compared to data from other rivers and established water quality standards.

This activity focuses primarily on identifying changes or trends in water quality over time. Students will investigate the interactions between several parameters in a river and the ways they affect each other. Specifically, they will notice the relationship between tempera-

ture and dissolved oxygen, and between flow and clarity. These relationships are discussed below:

Temperature: Soil erosion causes water to become murky, thereby increasing the absorption of the sun's rays. Thermal pollution—the addition of relatively warm water to a waterway from a power plant or urban runoff from warmed streets and pavement—can increase the temperature of a river.

Temperature has a direct influence on the amount of dissolved oxygen in a river. Generally, warmer waters have lower dissolved oxygen levels than colder waters. Further, aquatic organisms' metabolic rates increase in warm water, consuming even more oxygen from the water. An organism's ability to fight disease, parasites, or pollution is compromised when available dissolved oxygen decreases.

Dissolved Oxygen (DO): Most aquatic plants and animals, like humans, need oxygen to live. However, aquatic organisms need oxygen that is *dissolved* in water, and not atmospheric oxygen as humans do. Oxygen becomes dissolved in water in various ways, including tumbling over rocks (whitewater) or falls where water is agitated and atmospheric oxygen is forced into solution. Photosynthesis from aquatic plants also produces dissolved oxygen. While dissolved oxygen is retained in cool water, it is forced out of warmer water. The amount of oxygen dissolved in water peaks in late afternoon when photosynthesis is also at its peak and is lowest just before dawn.

With little agitation and typically higher temperatures, slower-moving, meandering rivers tend to have lower dissolved oxygen levels, while swift, well-aerated streams tend to have a high level of dissolved oxygen. In general, higher dissolved oxygen levels indicate healthier water for aquatic organisms.

Clarity: Clarity refers to how clear water is, or how deep one can see into the water. Relative clarity is measured in centimeters of depth of visibility. Clarity decreases with the increase in levels of suspended solids such as plankton, silt, clay, sewage, organic matter, and industrial waste. Suspended solids are introduced into watersheds by various means. Spring runoff, forest fires, and storms can all increase the rate of erosion and thus the amount of sediment that enters a waterway. Urban runoff, industrial effluent, and wastewater discharge can also contribute suspended solids to a watershed.

Water temperature increases as the clarity of a waterway decreases. When suspended solid levels increase so does the rate of absorption of the sun's rays. Clarity tends to be lowest following storms or during spring runoff when the water carries more eroded sediments and there is an increased disturbance of materials from the stream bottom.

Flow: A river flows at various rates throughout a watershed and throughout the seasons. Flow, or the speed at which water moves past a given point in one second, is typically measured in cubic feet per second or cfs. Both the width of a stream and the depth of its channel play a role in determining flow. Flow tends to be higher in the springtime when storms and snowmelt add volume to rivers. Depending on your geographic location, flows tend to be lower in the winter when much of the precipitation decreases or is frozen as snow.

Understanding the relationships between water quality parameters such as these and how these parameters can change throughout a watershed and over the seasons is an important first step toward understanding the broader scope of watershed monitoring.

Procedure

Warm Up

Discuss why water is monitored. Discuss the four parameters tested in this activity (temperature, dissolved oxygen, clarity, and flow), and the significance of each parameter. For example, why is it important to measure the temperature of water? Why should we be concerned about the rising temperature of a river? How will that affect the river and the aquatic organisms that inhabit the river?

The Activity

1. Distribute the **Cooper River Watershed Map** to each student.
2. **(Optional map-reading activity. If your students are skilled at reading topographic contour maps, please skip to Step 3.) Instruct students to locate the contour lines on the map. Contour lines are spaced at twenty-foot intervals and allow the reader to interpret what a three-dimensional landscape looks like using a two-dimensional map.** For example, contour lines allow the reader to determine the direction a river flows, where mountains and valleys and even waterfalls are located.

a. Have students place an X on the highest and lowest points on the map.

b. Have the students determine which direction the river is flowing (from highest to lowest elevations). Have them denote this by placing arrows on the map indicating the direction of flow.

c. Have students identify where on the river the waterfall is located. How did they identify this feature? (Two or more contour lines spaced tightly together indicate a very steep slope, or waterfall.)

3. **Drawing on what the students know about water flow, have them label the appropriate sampling sites with the following terms:** *Headwaters, Confluence, Midriver, Falls, Downriver.*

4. **Distribute a different Cooper River Water Data Card to each student.** Ask them to study their card and locate their site on the map.

5. **Distribute a copy of the** *Watershed Data Trends*

Worksheet to each student. Under the *Single Sample* section, ask students to write a brief summary of what they have learned about the water quality at their site. (They should only be able to state how the water at their monitoring site sample looked at that time, not any broader inferences about the river.)

6. Have selected students read their paragraphs and discuss the value of the information on the card. What other information about the river is needed to write a more complete paragraph?

7. Have students break into groups representing the five sections of the river, Headwaters, Confluence and so on through Midriver, Falls, and Downriver.

8. Instruct students to graph the data for their group. They should create a total of four graphs of data, one each for Temperature, Dissolved Oxygen, Clarity, and Flow. The parameter measured should be on the y-axis (vertical axis), and the seasons should be along the x-axis (horizontal axis). Their graphs should illustrate the seasonal temperature changes of the river at their site.

See more examples at www.healthywater.org.

9. Instruct each group to write a paragraph describing what they know about the entire river based on the data they graphed at their sampling site and then record the group paragraph under the *Sampling Site* section of their *Watershed Data Trends Worksheet.*

10. Ask groups to share their summaries with the class. What other information would they like to help them understand the water quality of Cooper River?

11. Reshuffle students into four groups representing each of the four seasons: winter, spring, summer, fall. All of the students with spring cards will form a "spring group" while all of the students with a winter sampling card will form a "winter group" and so on.

12. Have students graph each parameter along the y-axis and their sampling site along the x-

Waterfall at Horseshoe Rapids of the Chattahoochee River, Georgia. Courtesy: Joe and Monica Cook

axis. The groups should again have the same four graphs, one for each parameter, which allows them to interpret how each parameter changes as they follow the river down the watershed during their respective season.

13. Instruct the groups to write a paragraph describing what they know about the river based on the seasonal data they graphed. Record this paragraph under the *Seasons* section of the **Watershed Data Trends Worksheet.**

Wrap Up

Have students review their summary paragraphs and then read aloud selected ones to the class. Instruct students to write a final summary paragraph (under the *Project Summary* section) explaining how their site changed both spatially (along a watershed) and chronologically (through the seasons). What patterns or relationships were revealed from the spatial interpretation of the parameters? What patterns or relationships were revealed from the chronological interpretation of the parameters? Ask students to compare these summary paragraphs with their single sample paragraph based on their individual card.

Have students discuss the benefits of long-term data (a monitoring project) over a single monitoring event (a single sample). Why is it important to monitor a river over a long period of time? Discuss monitoring a river over time, and then monitoring an entire watershed over time. Which would be most valuable? What are

some limitations of both? (e.g., money, time, accessibility to each site, weather, etc.)

Assessment

Have students:
- identify important water quality parameters and their significance (**Warm Up**).
- read and interpret a contour map to determine the characteristics of a watershed (**Steps 2-3**).
- discern the value of data from a single sample and that from a series of samples of stream monitoring data (**Steps 4-13; Wrap Up**).
- graph, analyze, and summarize chronological and spatial water quality data for several parameters (**Steps 8-12**).
- compare and contrast the effects of water quality parameters on one another, especially the relationships between temperature and dissolved oxygen, and the relationships between flow and clarity (**Wrap Up**).

Extensions

Challenge students to use all the clues provided (temperature, elevation, etc.) to determine where this river may be located. Have them list their evidence and defend their prediction. The data is fabricated, but the data resembles that of a northern latitude state (Minnesota, Maine, Washington, Wisconsin) or Canadian province.

Research the watershed monitoring being conducted in your area. Invite a watershed monitoring coordinator or

water quality district representative to demonstrate water quality monitoring equipment and methods, or to discuss water quality issues in your area.

Resources

Ebbing, D. 1990. *General Chemistry, Third Edition.* Boston, MA: Houghton Mifflin Co.

Mitchell, M., and W. Stapp. 1997. *Field Manual for Water Quality Monitoring: An Environmental Education Program for Schools.* Dubuque, IA: Kendall/Hunt Publishing Co.

United States Department of Agriculture. 1991. *Water Quality Indicators Guide: Surface Waters.* Washington, D.C.: U.S. Government Printing Office.

United States Environmental Protection Agency. 1996. *The Volunteer Monitor's Guide to Quality Assurance Project Plans.* Washington, D.C.: U.S. Government Printing Office.

Yates, S. 1991. *Adopting a Stream: A Northwest Handbook.* Seattle, WA: University of Washington Press.

 Web sites: United States Environmental Protection Agency. *Monitoring Water Quality.* http://www.epa.gov/OWOW/monitoring/monintro.html

United States Geologic Survey. *National Water Quality Monitoring Council Work Plan.* http://water.usgs.gov/wicp/acwi/monitoring/nwqmc_wkpln.html

Cooper River Watershed Map

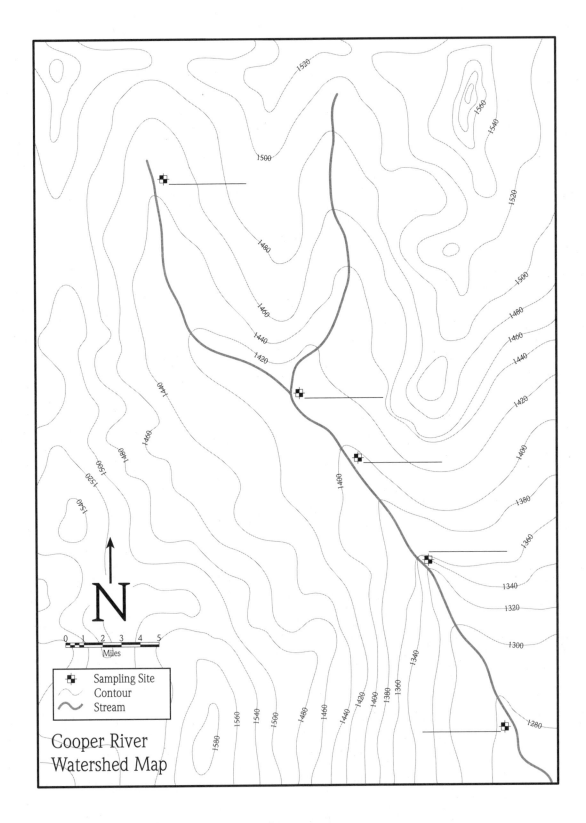

Cooper River Water Data Cards

Headwaters: Date: January 5 Conditions: River is partially frozen; clear day. Temp: .5° C DO: 14.5 mg/L Clarity: 48 cm Flow: 800 cfs	**Headwaters:** Date: April 7 Conditions: River is at high flows due to spring runoff. Temp: 4.4° C DO: 14.5 mg/L Clarity: 6 cm Flow: 2000 cfs	**Headwaters:** Date: July 6 Conditions: Hot and humid, but the trees over the river offer shade. Temp: 12.8° C DO: 12 mg/L Clarity: 26 cm Flow: 1200 cfs	**Headwaters:** Date: October 9 Conditions: Cool, fall day; leaves falling in the water. Temp: 10° C DO: 13 mg/L Clarity: 33 cm Flow: 750 cfs
Confluence: Date: January 5 Conditions: The combined flows keep the river free of ice. Temp: .5° C DO: 14.5 mg/L Clarity: 46 cm Flow: 1600 cfs	**Confluence:** Date: April 7 Conditions: Spring rains have led to high runoff. Temp: 4.4° C DO: 14.5 mg/L Clarity: 5 cm Flow: 4000 cfs	**Confluence:** Date: July 6 Conditions: Much of the bank is showing due to low flows. Temp: 11.7° C DO: 11.5 mg/L Clarity: 20 cm Flow: 2000 cfs	**Confluence:** Date: October 9 Conditions: Cool temperatures; leaves are changing colors. Temp: 9.4° C DO: 12 mg/L Clarity: 32 cm Flow: 1500 cfs
Midriver: Date: January 5 Conditions: Small ice sheets float by as you conduct your sampling. Temp: 1.7° C DO: 12 mg/L Clarity: 42 cm Flow: 1600 cfs	**Midriver:** Date: April 7 Conditions: Sediment levels are high due to the high flows. Temp: 7.2° C DO: 13 mg/L Clarity: 8 cm Flow: 4000 cfs	**Midriver:** Date: July 6 Conditions: Algae is beginning to grow along the edges of the river. Temp: 16° C DO: 8 mg/L Clarity: 17 cm Flow: 2000 cfs	**Midriver:** Date: October 9 Conditions: The water feels cool to the touch as fall approaches. Temp: 11° C DO: 9 mg/L Clarity: 24 cm Flow: 1500 cfs

Cooper River Water Data Cards

Falls: Date: January 5 Conditions: The churning water at the falls keeps ice from forming on the river. Temp: 2.8° C DO: 13 mg/L Clarity: 32 cm Flow: 1600 cfs	**Falls:** Date: April 7 Conditions: High flows have almost washed out the falls. Temp: 8.9° C DO: 12.2 mg/L Clarity: 8 cm Flow: 4000 cfs	**Falls:** Date: July 6 Conditions: Swimmers frequent the pool below the falls in summer. Temp: 18.3° C DO: 9 mg/L Clarity: 11 cm Flow: 2000 cfs	**Falls:** Date: October 9 Conditions: Low autumn flows greatly reduce the size of the falls. Temp: 12.8° C DO: 10 mg/L Clarity: 18 cm Flow: 1500 cfs
Downriver: Date: January 5 Conditions: Winter winds blow strong, keeping the river free of ice. Temp: 3.3° C DO: 6 mg/L Clarity: 32 cm Flow: 1600 cfs	**Downriver:** Date: April 7 Conditions: Brown, turbid water from high flows. Temp: 10° C DO: 7 mg/L Clarity: 8 cm Flow: 4000 cfs	**Downriver:** Date: July 6 Conditions: Dark, deep, and warm water slowly passes by your site. Temp: 22.8° C DO: 3 mg/L Clarity: 11 cm Flow: 2000 cfs	**Downriver:** Date: October 9 Conditions: The water seems clearer now than in the summer. Temp: 15° C DO: 4 mg/L Clarity: 18 cm Flow: 1500 cfs

Watershed Data Summaries Worksheet

Single Sample

Sampling Site

Seasons

Project Summary

Water Quality Monitoring:
From Design to Data

Summary

Students create a study design, then analyze and interpret water quality data to model the process of water quality monitoring.

Objectives

Students will:

- investigate the process of water monitoring.
- develop a study design that leads to a monitoring project.
- design an investigation to answer a research question.
- collect and analyze data and develop findings, conclusions, and recommendations from that data.

Materials

- *6 copies of **Table Monitoring Worksheet** Student Copy Page*
- *copy of **Table Monitoring Goals** Teacher Copy Page*
- *Ruler, tape measure, and graduated cylinders (optional) to measure changes in objects*
- *6 copies of **Gallatin Watershed Overview** Student Copy Page*
- *6 copies of **Gallatin Watershed Data Set** Student Copy Page*
- *6 copies of **Watershed Trends Worksheet** Student Copy Page*
- *Graphing paper (optional)*

Background:

Bodies of water are complex systems containing many measurable vari-

Grade Level:
9–12

Subject Areas:
Chemistry, Environmental Science, General Science

Duration:
Preparation:
20 minutes

Activity:
two-three 50-minute periods

Setting:
Classroom

Skills:
Gather, Analyze, Interpret, Draw Conclusions, Present

Vocabulary:
monitoring, indicator, parameter

ables. These variables can serve as indicators of water quality and watershed health. In a perfect world we would monitor *every measurable parameter,* at *all times,* and at *all points* of a watershed. Of course, this is impossible and impractical. Instead, decisions must be made about what parts of a complex system will be measured when and where measurements will be taken and by whom. Just as a dentist must decide among a myriad of variables when performing a routine checkup, so water quality monitors must choose just how to go about their work. Dentists generally look for indicators of decay, injury, and deterioration of gums and teeth. These check-ups are usually conducted at six-month intervals. In most situations, it would be impractical to check more often. The same is true in water quality monitoring. With so many variables and parameters as well as the complexity of water systems, how can water monitors decide just how, when, where and what to monitor? Who will analyze, interpret and present any data gleaned from such monitoring? The best way to answer these questions and make choices leading to a scientifically credible monitoring program is to develop a study design.

A study design is a document that describes decisions made about a monitoring program. Study design is also a process that allows people to ask essential questions about water quality monitoring. What can be gained from a monitoring program? What is appropriate data? Where and when will data be collected? Who will interpret the data? How will data be used to make decisions?

There are many reasons to develop a study design. Consider the following scenarios where a study design could meet each individual need.

- A school wants to monitor a stream after a water quality restoration plan has been approved. Teachers and students must determine what sampling protocols would give them appropriately accurate data.
- A Conservation District has diligently implemented new land use management practices designed to improve water quality within a watershed. Now, they want to find out if any of the changes actually improved water quality.
- A River Task Force is funding a monitoring program to assess the cumulative impacts of bank erosion. The Task Force Monitoring Specialist understands the data but will it make sense to the rest of the group?

- A school's water monitoring program is plagued by teacher turnover. The program loses momentum each time a teacher leaves because new teachers must decipher the goals and activities of the program without the aid of a written document.

Study design mirrors the scientific method in that it strives to answer a research question by designing a method of testing the question, collecting and analyzing data, drawing conclusions and offering recommendations from the analysis. Study design directly correlates with the investigative process of inquiry described by the National Research Council. "Inquiry is a multi-faceted activity that involves making observations; posing questions; examining books and other sources of information to see what is already known; planning investigations; reviewing what is already known in light of experimental evidence; using tools to gather, analyze, and interpret data; proposing answers, explanations, and predictions; and communicating the results. Inquiry requires identification of assumptions, use of critical and

logical thinking, and consideration of alternative explanations" (NRC, 2000).

The steps of a carefully planned study design are:

1. Define the <u>monitoring goal, or purpose</u>

Why are you monitoring? What water quality questions do you want to answer? For example, if the issue is the threat of pollution entering a stream from the mall parking lot, you might ask the question: *Is parking lot runoff adversely affecting aquatic life?* The definition of your monitoring purpose will influence the decisions you make on what, where, when and how to monitor.

2. Decide what will be monitored

What indicators or parameters will you monitor? How will these parameters attain your monitoring purpose?

3. Identify the <u>data quality objectives</u> (DQOs)

Who will use the data, and for what purpose? What level of precision and accuracy is required? DQOs can help determine specific methods. For instance, if you need to record pH levels to the tenths place, a pH strip won't work. You will need an electronic meter to provide the necessary sensitivity.

4. Choose how to monitor

What are your sampling and analysis procedures? Describe how samples will be collected, analyzed, and reported.

A volunteer water quality monitor sampling the Gallatin River, Montana. Courtesy: Gallatin Blue Water Task Force.

5. Decide where to monitor

Where are your monitoring sites? What are the criteria for choosing these sites? Include site description, latitude/longitude and rationale for choosing each site.

6. Decide when to monitor

What is your sampling frequency (once a year, biannually, monthly)? What time of day will you sample? List the sampling dates, times, and the rationale for choosing them.

7. Document who will do the monitoring

Who will monitor the data and what is their level of expertise? List the name, contact information, and responsibilities of the monitoring personnel.

8. Identify Quality Assurance (QA) measures

How will you document project procedures to ensure that the data collected meets data quality objectives?

9. Manage, analyze and report data

Managing data–How will data be recorded, entered and validated?

Analyzing data–How will data be interpreted?

• assemble information (maps, data sets, other)

• develop observations about the data

• develop conclusions about why the data look the way they do

• recommend action or further study

Reporting data–How will data be presented so different audiences can understand it?

A study design should precede any water quality investigation or field study. Study designs help monitoring groups focus on what they want to monitor, allow students to methodically investigate a question, and model a scientific investigation.

Procedure

Warm Up

1. Place numerous objects on your desk—water bottle, cup of coffee, pencils and pens, chalk, papers, open and closed books, picture frames, etc. There should be at least ten items on the desk, preferably many more. Explain to the students that they will individually monitor and record any changes in the objects on the table. Have them gather around the table and study it for 30 seconds, then return to their seats.

2. Begin a discussion about water monitoring. While you talk, alter the items on the desk (drink the liquids, turn book pages, use chalk and writing implements, etc.). Ask students if they have ever been to a dentist? Have them brainstorm the variables, or parameters, that the dentist is checking or *monitoring* during a dental exam. Examples include teeth hardness, whiteness or color, plaque build-up, cavities, alignment, etc. Ask students why they don't have a dental exam every day. (Don't their teeth change a little bit every day, especially if there is a cavity?) Since it is not practical to have a dental exam every day—the dentist instead monitors teeth at six-month intervals, looking for indicators or problems of declining oral health.

3. Describe how monitoring a stream can be similar to a dental exam. To determine the health of a body of water, monitoring _all parameters_ at _all points_ at _all times_ is important. However, like a daily dental check up, it is impractical, if not impossible to monitor *everything, everywhere, all the time.* Discuss how choices must be made in order to conduct a scientifically credible monitoring program. Ask students to name other things that undergo periodic monitoring—drinking water, blood pressure, weather-related parameters, economy, etc.

4. Commonly, water quality monitoring involves repeatedly testing water for specific water quality variables (pH, dissolved oxygen, turbidity, etc.). Ask students what variables or parameters of a river might change, and which are indicators of a water body's health. Examples include temperature, pH, dissolved oxygen, turbidity, flow, bacteria, aquatic insects, alkalinity, heavy metals, etc. Write these parameters on the board. Ask students why each of these need monitoring.

Gallatin Watershed Maps

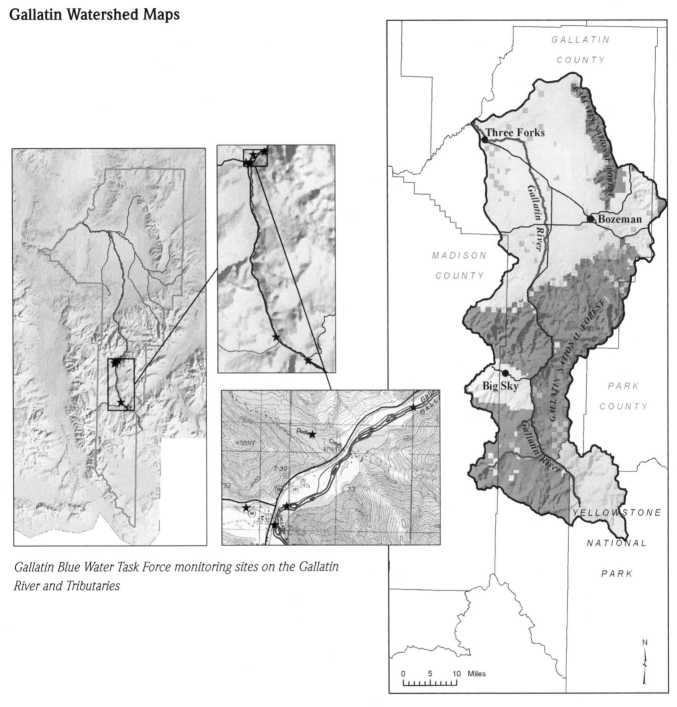

Gallatin Blue Water Task Force monitoring sites on the Gallatin River and Tributaries

Gallatin Watershed

Maps provided by Katie Alvin of Carto-Logic, and the Gallatin Blue Water Task Force.

5. After the discussion, have students look at the objects on the desk again and record which objects changed and how. Was each student able to record ALL of the changes? Why not? (There were so many variables to monitor that they couldn't keep track of them all.) Explain that in the next part of this activity, they will use a study design to choose specific parameters to monitor.

The Activity

Part I

1. **Divide students into six groups and distribute the *Table Moni toring Worksheet* to each group.** Their task is to "monitor" items on your desk or table at 5-minute intervals for ten minutes. Place numerous items on your desk; include some that change and some that don't (as in the ***Warm Up***). Add more objects to increase variety. Review the ***Table Monitoring Goals*** to ensure the objects on the table meet these goals.

2. **Before monitoring, explain that all monitoring projects must start with a target or purpose.** Copy the ***Table Monitoring Goals*** and distribute one card to each group. Have students focus on items specific to their assigned monitoring goal. Have them write their goal in the space provided on the ***Table Monitoring Worksheet.***

3. **Groups must decide which parameters or objects they will monitor in order to meet their goal.** Have them list their parameters in the space provided on the ***Table Monitoring Worksheet.***

4. **Instruct groups to record baseline data on their parameters.** Baseline data is an initial data set that will be compared to subsequent data so changes can be noted. To establish a baseline, students should record how their chosen parameters look at the start. They must also decide how they plan to record their data (e.g., draw a picture of the object or write a narrative description). Give them ample time to gather this baseline data. When all groups have recorded their baseline data, explain that they will monitor this same desk in five minutes and again in ten minutes, and record any changes they observe in their chosen parameters on their ***Table Monitoring Worksheets.***

5. **Start timing. While students share their monitoring goals and parameters with the class, casually alter the items on the desk (e.g., drink the coffee, turn pages in a book, use chalk to write on the board, etc.).** Another option is to have the table in a separate room where an assistant makes the changes to the objects. The students then revisit the table as if they were visiting a stream.

6. **After five minutes, have groups observe the items and record their observations (data) as before.** Allow the groups an appropriate amount of time to collect their data, but not too much.

7. **Ask groups to share what they observed after the first monitoring session (while again changing items on the table).** After five more minutes, have the students record their observations again.

8. **Instruct the groups to analyze their data and record any changes that they noticed in their parameters.** Lead a discussion by asking the following questions:
   - What changes did you notice?
   - What trends did you observe in your data? (e.g., there was less coffee in the mug each time; the clock continued to move in the same direction at a steady rate.)
   - What parameters stayed the same?

Part II

1. **Review the steps conducted in *Part I*:**
   a. Establish a monitoring goal
   b. Choose parameters to meet your monitoring goal
   c. Collect and record meaningful data
   d. Analyze the data by:
      1. gathering all needed information.
      2. listing results or findings of the data.

The next two steps in the process are : (not used in *Part I*)

3. develop conclusions based on findings.

4. present findings in an organized way.

2. Explain to students that they will now expand this study design to analyze and interpret actual water quality data from a watershed.

3. Distribute the *Watershed Data Worksheet, Gallatin* Watershed Overview, and the Gallatin Watershed Data Set to the six groups.

4. Have groups write a paragraph describing the land uses occurring in the watershed and the *possible* water quality impacts from them. (Students may conduct further research on the Gallatin River watershed and land uses at this time, or select one of the other *Watershed Data Sets* for a watershed closer to their region of the country.)

5. Have students refer to the *Gallatin Watershed Overview* to answer the questions on their *Watershed Data Worksheet*.

6. Assign one parameter to each group. Groups should examine the data for all the testing sites and dates to look for trends and anomalies (data that is out of line with the expected trends). Graphing data effectively illustrates trends and anomalies in water quality data.

7. Have groups answer the following questions:
What changes did you notice? What trends did you observe in your data?

What parameters stayed the same over the monitoring period?

8. Have groups write a paragraph explaining their conclusions, or rationale for their findings. In real-life monitoring, the actual cause of anomalous data is often unclear. Causes may be attributed to testing error (Was it the middle of winter when sampling accuracy is affected by freezing temperatures?) or perhaps a weather event that led to unusually high or low results.

9. Have groups present a summary of their data, findings, and conclusions to the class. If time allows, encourage groups to use PowerPoint or another creative presentation media.

Wrap Up

Have each group compare their parameter data with parameters set by other groups. Identify any relationships between the parameters (e.g., dissolved oxygen goes down as temperature goes up). Have students record their results (e.g., make overheads of their graphs and overlay them with other graphs to analyze relationships between the parameters). The Analysis Findings below were written by the Gallatin Blue Water Task Force and reflect a synopsis of possible explanations for variances in the data. Share these with students after they have presented their findings.

Discuss how difficult it is to draw specific conclusions about the cause of water quality changes in this watershed based on limited data. Explain that this is a common challenge, as the exact cause of a change in water quality is frequently unknown. This uncertainty is the reason that water quality monitoring is critical. What findings and conclusions can be drawn from one sample data set from one date only?

Assessment

Have students:

- investigate the process of water monitoring (*Parts I and II*).
- develop a study design that leads to a monitoring project (*Parts I and II*).
- design an investigation to answer a research question (*Parts I and II*).
- collect and analyze data; develop findings, conclusions, and recommendations from that data (*Part II* and *Wrap Up*).

Extensions

Students can plan and implement their own water quality monitoring project to study aspects of a local waterway. Have them design the project so that subsequent classes can continue to collect data over several years to allow trends in the data to become more apparent.

Conduct a "town meeting" using the students' findings and conclusions. Have groups of students represent different stakeholders in the meeting —including a monitoring group to share the class findings at the meeting. The other groups could

Analysis Findings from the Gallatin Blue Water Task Force:

a. The Taylor Fork drainage has many raw cliffs adjacent to the stream that erode during spring runoff and after precipitation events. The erosion is severe, and the Gallatin River becomes dull gray in color after these events. There is debate as to whether the erosion is natural or human-induced. There has been extensive logging in the Taylor Fork drainage, but the soils in the area are also susceptible to erosion. This is evidenced by the turbidity data—the highest readings after spring runoff are at the Taylor Fork site, and the readings decrease farther downstream from the Taylor Fork. During this time, turbidity in the West Fork also is slightly increased. There is a lot of development within the West Fork drainage that may contribute to increased sediment loads in the stream.

b. Dissolved oxygen readings in December 2000 at some sites are extremely low. This probably does not reflect actual DO levels in the stream, but rather a malfunction of equipment in the field. Chemical reactions may have been inhibited by the cold weather that day (-20F).

c. At two sites, Upstream and Downstream I, several springs drain into the main channel of the river. This may slightly affect some water quality parameters at these sites (i.e. temperatures may be slightly warmer at these sites during winter months).

d. The most-impacted site is the West Fork of the Gallatin. Most of the development in the Big Sky area is taking place in the West Fork drainage. Analyses of aquatic insects determined this site was slightly impaired; all other sites were unimpaired.

Resources

Dates, G. 1995. *Study Design Workbook.* Montpelier, VT: River Watch Network.

Mayio, A. 1997. *Volunteer Stream Monitoring: A Methods Manual.* Washington, D.C.: United States Environmental Protection Agency.

National Research Council. 2000. *Inquiry and the National Science Education Standards.* Washington, DC: National Academy Press.

River Network and Pennsylvania Department of Environmental Protection, Bureau of Watershed Conservation, Citizens' Volunteer Monitoring Program. 2000. *Designing Your Monitoring Program: A Technical Handbook for Community-Based Monitoring in Pennsylvania.* Harrisburg, PA: Pennsylvania Department of Environmental Protection.

represent local landowners, farmers, businesses, etc. Have each group react to the findings and recommendations of the monitoring group.

Have students research water quality monitoring data sets from local and state government agencies or water monitoring groups. Have them analyze the data using the process from *Part II*.

Testing Kit Extensions
Develop a study design for a water quality monitoring project on a local stream or lake. Use the *Healthy Water, Healthy People Testing Kits* to conduct water quality tests to fit chosen parameters.

Table Monitoring Goals

Group 1
Monitoring Goal:
Interested in tracking the changes in the amounts of all liquids.

What you will monitor:

Group 2
Monitoring Goal:
Interested in tracking changes in objects that are taller than 5 cm.

What you will monitor:

Group 3
Monitoring Goal:
Interested in tracking changes in the size and location of writing implements.

What you will monitor:

Group 4
Monitoring Goal:
Interested in tracking changes in objects that are primarily blue, red, and green.

What you will monitor:

Group 5
Monitoring Goal:
Interested in tracking changes in objects that are primarily white, black, and grey.

What you will monitor:

Group 6
Monitoring Goal:
Interested in tracking changes in objects that are greater than 50 cm^2 in size.

What you will monitor:

Table Monitoring Worksheet

1. Monitoring Goal (from monitoring goal card):

2. What parameters you will monitor (from monitoring goal card):

3. Data Collection: Baseline Data—Draw or note the objects you are monitoring as they exist in their original state or position.

After 5 minutes—Draw or note what you observe about the objects you are monitoring.

After 10 minutes—Draw or note what you observe about the objects you are monitoring.

4. Data Analysis and Findings:
   - What changes did you notice?
   - What trends did you observe in your data? (e.g., was there less coffee in the mug each time?)
   - What parameters stayed the same over the monitoring period?

Watershed Trends Worksheet

Refer to the *Gallatin Watershed Overview* and *Gallatin Watershed Data Set* to complete the following study design:

1. Monitoring Purpose:

2. Parameters to be monitored (see *Gallatin Watershed Data Set*):

3. Data Quality Objectives:

4. How monitoring will be conducted:

5. Where monitoring will be conducted:

6. When monitoring will be conducted:

7. Who will conduct monitoring:

8. Quality Assurance Measures:

9. Data Analysis: Develop Findings about the Data:
   a. Graph the data for your parameter. *{See examples @ www.healthywater.org}*
      I. What trends do you observe?
      II. Do results change upstream to downstream?
      III. Where in the watershed did you notice changes in data?
   b. Develop conclusions based on your findings about the data?
      a.
      b.
      c.
   c. Develop recommendations (e.g., no action, ways to reduce pollution, further monitoring, etc.):
      a.
      b.
      c.

Gallatin Watershed Data Set

Dates	Streams	pH (°C)	Water Temp (FTU)	Turbidity (mg/L)	Nitrate (mg/L)	D.O.	Percent Saturation
7-29-00	Park Boundary	8.3	8.0	——	0.02	10	85
	Upstream	8.1	11.0	——	0.05	8	80
	West Fork	8.4	10.0	——	0.04	10	90
	Downstream 1	7.4	11.0	——	0.04	9	82
	Dudley Cr.	7.7	6.5	——	0.06	11	90
	Downstream 2	8.6	12.5	——	0.03	10	100
9-9-00	Park Boundary	8.3	2.0	——	0.03	10.5	74
	Upstream	7.8	8.0	——	0.075	5	44
	West Fork	8.6	4.0	——	0.035	9.5	72
	Downstream 1	8.1	7.0	——	0.05	8	66
	Dudley Cr.	8.0	3.5	——	0.01	10	74
	Downstream 2	8.6	8.0	——	0.025	10	85
10-22-00	Park Boundary	7.5	1.0	——	0.045	9.5	67.5
	Upstream	7.9	0.0	——	0.04	9	62
	West Fork	7.6	-4.0	——	0.06	10.5	66.5
	Downstream 1	7.9	5.0	——	0.05	9	70
	Dudley Cr.	8.0	-2.0	——	0.04	11	73
	Downstream 2	8.6	1.0	——	0.03	10.5	72.5
12-2-00	Park Boundary	7.7	0.0	——	0.085	2.3	15
	Upstream	7.1	7.0	——	0.09	2.75	24
	West Fork	7.8	1.0	——	0.19	1.5	11
	Downstream 1	7.8	7.0	——	0.12	1.5	13
	Dudley Cr.	7.8	1.0	——	0.09	7.75	55
	Downstream 2	8.1	4.0	——	0.06	10.25	77
1-13-01	Park Boundary	7.5	-2.0	——	0.085	11	70
	Upstream	7.4	0.0	——	0.075	9	61
	West Fork	7.4	-2.0	——	0.185	11.5	73.5
	Downstream 1	7.5	3.0	——	0.12	9	67
	Dudley Cr.	8.0	-2.0	——	0.05	11.5	73.5
	Downstream 2	8.3	3.0	——	0.075	10	74
3-24-01	Park Boundary	8.0	3.0	——	0.045	12	87
	Upstream	7.8	3.5	——	0.04	11	80
	West Fork	8.3	2.0	——	0.125	12	85
	Downstream 1	8.0	2.5	——	0.055	8.5	61.5
	Dudley Cr.	8.0	4.0	——	0.07	11.5	86
	Downstream 2	8.4	7.0	——	0.025	10	82
4-28-01	Park Boundary	7.6	4.0	——	0.04	11.36	85
	Upstream	7.5	9.0	——		11.4	97
	West Fork	7.2	4.0	——	0.085	11.9	89
	Downstream 1	7.6	5.0	——	0.075	11.4	87
	Dudley Cr.	7.6	7.0	——	0.08	11	90
	Downstream 2		6.6	——		10.2	82

Dates	Streams	pH	Water Temp (°C) (FTU)	Turbidity (mg/L)	Nitrate (mg/L)	D.O.	Percent Saturation
7-14-01	Park Boundary	8.1	10.9	4	0.035	9.5	86
	Taylor Fork	7.9	12.3	7	0.035	10	95
	Upstream	8.3	15.1	7.5	0.035	9.5	95
	West Fork	8.7	13.2	7.5	0.025	10.5	100.5
	Downstream 1	8.5	14.0	6.0	0.035	10.5	102
	Dudley Cr.	8.2	13.1	1.0	0.04	10	96
	Downstream 2	8.5	16.4	6.5	0.03	9.5	99.5
8-12-01	Park Boundary	7.6	8.0	2.5	0.025	11.5	96
	Taylor Fork	8.2	9.0	3.5	0.025	9.5	82.5
	Upstream	8.0	13.0	0.0	0.03	9.5	91.5
	West Fork	8.4	12.2	2.5	0.03	10	95
	Downstream 1	8.4	13.9	1.0	0.045	9	88.5
	Dudley Cr.	8.3	12.2	1.5	0.05	9.5	89.5
	Downstream 2	8.6	15.5	2.0	0.03	9.5	96
9-15-01	Park Boundary	8.0	5.3	0.5	0.03	12	93
	Taylor Fork	8.4	6.7	0.5	0.02	10	80
	Upstream	7.2	10.0	1.5	0.04	9.5	84.5
	West Fork	8.5	8.2	3.0	0.05	10.5	88
	Downstream 1	8.3	10.9	1.5	0.06	9	82
	Dudley Cr.	7.9	8.8	0.5	0.045	10	85
	Downstream 2	8.6	12.5	0.5	0.03	9.5	90
10-13-01	Park Boundary	7.1	0.0	0.0	0.05	12.5	83.5
	Taylor Fork	7.2	1.0	0.5	0.05	12	83
	Upstream	7.3	2.8	1.0	0.04	12	85
	West Fork	8.0	1.4	1.5	0.1	12.5	84.5
	Downstream 1	8.0	4.7	0.0	0.06	11	85
	Dudley Cr.	7.9	2.4	0.5	0.055	12	85
	Downstream 2	8.2	5.8	1.0	0.04	11.5	90.5
11-10-01	Park Boundary	7.2	0.2	2.0	0.065	12.5	83.5
	Taylor Fork	7.5	0.0	1.5	0.05	13	85
	Upstream	7.8	1.8	3.0	0.05	8	57.5
	West Fork	8.0	0.1	2.5	0.165	13	87
	Downstream 1	7.8	6.3	2.0	0.12	9.5	77
	Dudley Cr.	7.8	0.7	1.5	0.1	12	80
	Downstream 2	8.1	5.0	1.5	0.05	13.5	102
12-8-01	Park Boundary	7.4	-0.3	1.0	0.11	14	90
	Taylor Fork	7.8	2.4	1.25	0.08	13.75	96
	Upstream	7.4	2.9	1.5	0.1	11	80
	West Fork	7.7	-0.4	1.5	0.26	12	79
	South Fork	8.0	-0.2	3.0	0.145	12	79
	Downstream 1	7.8	4.5	0.5	0.155	10	78
	Dudley Cr.	7.8	0.2	0.0	0.11	12	80
	Downstream 2	8.2	3.8	1.5	0.08	12	89

Gallatin Watershed Overview

Vital Statistics:

From its headwaters in Yellowstone National Park, located in southwest Montana, the Gallatin River flows northward through the Gallatin Canyon to Three Forks, Montana, where it joins the Madison and Jefferson Rivers to form the headwaters of the Missouri River. The river was named by the Lewis and Clark Expedition on July 25, 1805, after President Thomas Jefferson's Secretary of the Treasury Albert Gallatin. Along its length, the Gallatin River flows swiftly through Yellowstone National Park and US Forest Service Land before entering the predominantly private agricultural and urban landscape of the Gallatin Valley. The land immediately adjacent to the river and along Montana Highway 191 is private and used for residential homes, guest ranches, agriculture, ranching, and various tourist-related businesses including motels, restaurants, and gas stations. There is a large ski resort at the upper end of the West Fork, along with a community of residents and numerous ski and tourist-related businesses and accommodations.

Famous for its blue ribbon trout fishery and frequented by anglers from around the world, the river also accommodates rafters, hikers, campers, and sightseers on a year-round basis. In addition to the people that frequent this natural resource, the river corridor also provides habitat for a wide variety of wildlife: bear, moose, elk, deer, migrant and resident bird species, and native trout species, most notably cutthroat trout.

Population growth, however, is placing great demands on many of Montana's waters, and increases the potential for degradation of their clean and clear character. The Gallatin Blue Water Task Force was established in the spring of 2000 as a volunteer water quality monitoring group. Since that time, the Task Force has collected monthly water quality information at seven locations along the Upper Gallatin River and two of its tributaries near the ski resort town of Big Sky, Montana. Two monitoring sites (Park Boundary and Dudley Creek) were chosen as reference sites because human impacts to water quality should be minimal there. The other sites were chosen to measure potential human impacts to water quality in the Big Sky area over a long period of time.

The goals of the Blue Water Task Force are to collect rigorous water quality information every month for the Upper Gallatin River and to provide educational information to the public. The impetus for starting the program was concern about the lack of baseline water quality information and the potential for population growth to impact local water quality. Citizen volunteers, trained by the Blue Water Task Force, collect water quality information the second Saturday of every month. Data collected by the Task Force meet quality controls specified by the Montana Department of Environmental Quality. Equipment used in the field is calibrated before each collection day, and EPA-approved procedures and proper standards and controls are used for all laboratory analyses. Information collected each month by Task Force volunteers represents a snapshot of the river water quality at the time the sample was collected. Current sampling protocols are not designed to record daily events that may affect water quality but rather long-term trends in water quality over time.

Turbidity or Not Turbidity
That is the Question!

Summary
Students explore the effects of sediment on turbidity; compare the turbidity of muddy and clear water; simulate environmental conditions that cause erosion; and investigate ways to reduce erosion that leads to turbidity in adjacent waterways.

Objectives
Students will:
- explore the relationship of soil erosion to the turbidity of water.
- compare the turbidity of muddy and clear water.
- simulate environmental conditions that can cause high turbidity in rivers.
- simulate restoration efforts that reduce erosion and turbidity.

Materials
- *Clear plastic (or glass) quart jar filled with assorted rocks, gravel, soil, sand, and water*
- *Copies of **Turbidity Test** Student Copy Page (one per group)*
- *Turbidity test materials*
- *Flat-bottomed test tubes (preferred) or clear juice glasses (one per group)*
- *Fine-grained soil (e.g., silt) or milk (see **Warm Up: Part II – Preparation** for amounts to make turbid water)*
- *Sample of clear water (e.g., tap water–at least 1 liter)*

Grade Level:
6–8

Subject Areas:
Environmental Science, Math, Earth Science

Duration:
Preparation:
Setup 30 minutes

Activity:
50 minutes

Setting:
Outdoors or large gymnasium

Skills:
Measure, Interpret, Whole-body Movement, Compare and Contrast, Investigate, Problem-solve

Vocabulary:
turbidity, erosion, sediment, runoff, remediate, nonpoint source pollution, best management practice

- *Sample of local surface water for comparison (at least 1 liter)*
- *Optional: eyedropper, coffee filter*
- *Rope or string (about 15 meters)*
- *200 tokens (acorns, leaves, twigs, packaging peanuts, balls, other)*
- *Pencils*
- *Paper*

Background
Erosion increases the amount of sediment (soil particles) in water. This increased sediment influences turbidity, which is "an optical property of water based on the amount of light reflected by suspended particles" (USEPA, 1999). Thus, very turbid water appears murky or cloudy. All natural waters are somewhat turbid, even if only at microscopic levels.

Measuring turbidity, which will give an idea of the volume of suspended and colloidal matter present in a body of water at a particular time, can be one indicator in assessing water quality. Suspended and colloidal matter (microscopic particles that remain suspended in water and diffract light) can be anything that is suspended in the water column ranging from sand, silt, clay, plankton, industrial wastes, sewage, lead, and asbestos to bacteria and viruses. Some suspended matter occurs naturally and some is produced by human activities.

Aquatic organisms are particularly susceptible to the effects of increased sediments and turbidity. Many fish need clear water to spot their prey.

Macroinvertebrates, fish eggs, and larvae require oxygen-rich water circulating through clean gravel beds to survive. Sediments can smother fish eggs and aquatic insects on the bottom and can even suffocate clams and oysters as they filter water through their bodies. Sediment and other dissolved substances also decrease light penetration, which inhibits aquatic plant photosynthesis.

Because cloudy water absorbs more of the sun's energy than clear water, high turbidity also leads to higher water temperatures. This can severely affect aquatic organisms, many of which are adapted to survive within narrow temperature ranges.

Sediment and turbidity also affect people. According to the USEPA, sediment is one of the first things filtered out of source water at a drinking water treatment plant and is one of the few water quality contaminants that must be monitored daily. Suspended solids can harbor harmful bacteria and can also decrease the effectiveness of chlorination used to help remove those harmful bacteria. Increased turbidity and sediments have economic impacts as well. The production of everyday products, like paper, food, and computer chips, requires water that is free of sediment and other suspended solids.

The most common natural source of suspended matter is sediment washed by erosion and runoff into a body of water. Sediment carried

over land during runoff is the most common form of nonpoint source pollution in the United States. Human-caused erosion stems mainly from activities like road building, construction, agriculture, logging, and other endeavors that remove or disturb vegetation. Natural erosion and the resulting turbidity is a common occurrence in some watersheds.

Many rivers carry naturally high sediment loads due to the nature of the rock and soil layers that they pass through. Highly erodable rock layers, such as limestone and sandstone, are evidenced by turbid, muddy flows in the adjacent river. The Missouri River, also called the "Big Muddy," is an example of a river that is naturally turbid year-

round from the highly erodable sedimentary limestone that it courses through. The Colorado River rushes through sandstone canyons, like the Grand Canyon, and the highly turbid waters found there evidence these highly erodable rocks.

While erosion and turbidity adversely affect some aquatic organisms, several have adaptations that allow them to thrive in highly turbid waters. Paddlefish (*Polyodon spathula*), found in the upper Missouri River, have long, sensitive snouts that allow them to probe for food in the dark, murky water, while channel catfish (*Ictalurus punctatus*) have barbels on their snouts that serve the same function.

A healthy stream with little erosion and sediment due to the vegetative cover along the banks. Courtesy: U.S. Fish and Wildlife Service.

Best Management Practices to reduce erosion.	
Best Management Practice	**Explanation**
Buffer Zone	Involves leaving a natural, undisturbed strip or "greenbelt" on the surrounding an erosive area.
Constructed Wetland	A modified natural or constructed shallow basin for treatment of contaminated waters by wetland vegetation.
Impervious Surface Reduction	Reducing the amount of paved land reduces water velocities and erosion.
Mulches, Blankets, and Mats	Covering the soil with these protects the soil surface, reduces runoff velocity, increases infiltration, slows soil moisture loss, and improves seed germination.
Riprap	Rocks placed along a streambank to protect the soil from direct contact with erosive stream flows.
Sand Fence	A low fence of wooden slats erected perpendicular to the prevailing wind. Prevents wind erosion of sand.
Sediment Basin	A low earthen dam across a drainage way to form a temporary sediment storage pool.
Silt Fence	A permeable barrier designed to capture sediment from disturbed areas.
Straw Bale Barrier	A temporary sediment barrier of anchored straw bales.
Trees, Shrubs, Vines, and Ground Covers	If planted or left intact, can provide superior, low-maintenance, long-term erosion protection.
Vegetative Streambank Stabilization	Planting trees or shrubs along streambanks to stabilize them.

If erosion and runoff are the largest contributors to sediments in our waterways, then slowing runoff and mitigating erosion is crucial to improving water quality. Sometimes human-caused erosion can be managed to reduce the turbidity of waterways. The simplest way is to maintain or plant vegetation along the edge of waterways to serve as a buffer strip, capturing the sediments before they enter the water.

Plants slow the velocity of water as it runs off over the surface of the land. When runoff is slowed, erosion is decreased. Plants prevent soil from being washed away by holding it together. They also reduce the velocity at which raindrops hit the soil. When unobstructed raindrops strike the ground, they promote rapid runoff by forcing finer soil particles to the surface. These particles eventually create a barrier, which prevents water from soaking easily into the ground (Farthing et al., 1992).

Plants help store water by adding organic matter to the soil, which eventually decomposes into highly absorbent humus. Fallen leaves and stems also slow water down, giving it more time to soak into the ground. Roots also store water by conducting it away from the surface and into the ground.

Procedure
Warm Up

Part I
Fill a clear quart jar one-third full

with assorted rocks, pebbles, mud, sand, and soil. Add water until the jar is full. Shake the container to stir up the rocks, water, and soil until the water becomes muddy. Show students the sample of muddy water and ask if they can see through it. Would they drink the muddy water? Why not? How does cloudy water affect aquatic organisms?

As the mud, or sediment, in the jar settles, the heaviest rocks will be on the bottom, the smaller rocks above them, and the finer soil particles, or sediment, on top of the rocks covering them like a blanket. Have students brainstorm possible impacts of this sediment covering the rocks (e.g., can smother fish eggs, suffocate aquatic insects and plants, abrade fish gills and mollusks). Many fish have adapted to live in highly turbid waters, such as paddlefish and catfish. What adaptations do these fish have that allow them to thrive in turbid water?

Since the clarity of the water does affect living organisms, scientists have developed a turbidity test to measure water clarity. Explain to students that they will conduct a relative turbidity test using common materials.

Part II
Preparation: Prior to conducting **Part II**, premix a turbid solution (muddy water) using fine-grained soil (i.e. silt) so that soil stays suspended. (Milk, a substitute, creates a turbid solution of a

different color for comparison.)

Conduct the turbidity test on the **Turbidity Test** *Student Copy Page* ahead of time to ensure that your solution is neither too muddy nor too clear—it should test out at grayscale circles D or E with about one inch of solution in the tube or glass. If mud or heavy soil is used, pour the sample through a coffee filter to strain out the larger particles that could settle out and cover the bottom of the test tube, giving an inaccurate reading. If possible, distribute the solution to the students via an eyedropper, as this will ensure that *suspended* and not heavier particles are being used.
1. Have students work in cooperative groups. Distribute a test tube or juice glass and a *Turbidity Test* to each group.
2. Have students follow the testing procedures on the *Turbidity Test* recording the corresponding letter of their relative turbidity reading and its interpretation. Have students continually swirl their samples to ensure the sediment remains suspended and not settled on the bottom.
3. Repeat the turbidity test with clear water.
4. Repeat the turbidity test with the sample of local surface water.

Ask students to compare their results. What can account for the differences in turbidity between the solutions?

The Activity
1. **Prior to the activity, delin-**

eate a hillside or an area that will simulate a bare hillside— ap-proximately 15 meters wide. Use a rope to signify the edge of a river that runs along the bottom of the hill.
2. **Randomly scatter the tokens across the area.**
3. **Divide students into four groups, explaining that they will be simulating soil erosion, or sediment, that leads to high turbidity in the river below.**
4. **Assign three of the four groups of students to be water.** The fourth group will serve as data recorders for the first part of the activity, and plants in the second part.
5. **Have the "water" students stand side by side, facing down-hill, at the top of the slope or playing area.**
6. **Announce the arrival of a rainstorm, which prompts the water students to proceed down-hill.** As they walk, they should collect tokens, which represent sediment that is carried by the water as it moves down the hill. Remind students that water droplets do not wander from side to side, but rather pick up only the soil particles that lie in their path.
7. **Students continue down the slope until they congregate in the river at the bottom of the hill. Here they count the number of tokens (sediment) that reached the river.** The fourth group of students now records the number of tokens that each water student carried down the slope and deposited in the river. *(Option:*

Have students create graphs of the amount of sediment tokens present in the river after each round.)

8. Redistribute the tokens throughout the hillside.

9. Conduct the activity again, only this time instruct the fourth group of students to become plants that have been planted randomly on the bare hillside. The plant students stand throughout the slope with their arms out-stretched to represent roots.

10. Again, have the water students stand side by side at the top of the slope while you announce the arrival of a rain-storm, prompting them to proceed downhill. They will collect tokens as they walk down-hill. However, when the water students encounter a plant, they must circle that plant five times representing water being slowed

down by the plant. In addition, on every circle around the plant, they must drop a soil token demonstrat-ing how plants hold sediment on a hillside.

11. Once all of the water droplets have completed their journey into the river, they should count their tokens. Plants should record the number of tokens the water droplets carried down the slope and into the river during this round of the activity. (Option: Conduct the activity again, having the students strategize where to place the plants to most effectively decrease the sediment that reaches the stream. Have students research erosion control measures and add sediment fences and other erosion control measures to this activity. Play several more rounds allowing the students to compare the sed-iment levels caused by different

configurations of plant locations and erosion control measures.)

Wrap Up

After the students have completed the activity, discuss the results with them. Were there more soil particles in the river with or without vegeta-tion? In which round would the turbidity of the stream be higher—with or without vegetation? Why? What are some other ways to decrease the amount of soil erosion that enters a river? (Sediment fences, tree plantings, etc.) Why is it important that we try to reduce erosion and sediment in our streams?

Assessment

Have students:

- explore the relationship of soil erosion to the turbidity of waterways (***Warm Up–Part I).***
- compare water samples with high and low turbidity (***Warm Up–Part II).***
- quantify the turbidity of water samples using a turbidity test (***Warm Up–Part II).***
- simulate environmental condi-tions for high turbidity (**Steps 1-7**).
- simulate the remediation of environmental conditions for high turbidity (**Steps 8-11***).*
- compare the amount of sed-iment carried to the river before and after remediation (***Wrap Up).***

Extensions

Have students investigate the ranges

Erosion on a bare hillside. Courtesy: Natural Resurces Conservation Service

of turbidity in which aquatic organisms can survive. Are some species more tolerant to turbidity than others?

Contact your County Conservation District and ask their soil or water quality specialist to demonstrate the use of instruments that measure turbidity. What are some of the natural and unnatural contributors to turbidity that they have encountered? How do they suggest these contributors be remedied?

Survey your school grounds or a local park for evidence of erosion. Discuss ways that the erosion could be reduced or prevented. Working with officials from the school or park, help implement erosion control measures to reduce the erosion.

Testing Kit Extensions
Conduct the **Warm Up–**

Part II using a turbidity tube, Secchi disk, turbidimeter, or other instrument to monitor the trends in turbidity levels throughout the year for a local waterway. Compare these results with turbidity readings throughout the year and with other bodies of water.

Resources

Barzilay, J.I., J.W. Eley, and W.G. Weinberg. 1999. *The Water We Drink.* New Brunswick, NJ: Rutgers University Press.

Farthing, P., B. Hastie, S. Weston, and D. Wolf. 1992. *The Stream Scene.* Portland, OR: Aquatic Education Program.

Freedman, B. 1989. *Environmental Ecology.* San Diego, CA: Academic Press, Inc.

Important Water Quality Factors. 2001. Retrieved on September 4, 2001 from the Hach Company Web site: http://www.hach.com/h20u/h2wtrqual.htm

Murdock, T., & M. Cheo. 1996. *Streamkeepers Field Guide.* Everett, WA: The Adopt-A-Stream Foundation.

United States Environmental Protection Agency (EPA). 1999. *Guidance Manual for Compliance with the Interim Enhanced Surface Water Treatment Rule: Turbidity Provisions.* Retrieved on December 3, 2001, from the Web site: http://www.epa.gov/safewater/mdbp/mdbptg.html

Water Quality Index: Turbidity. 1999. Retrieved on October 8, 2001 from the Kansas Collaborative Research Network (KanCRN) Web site: http://www.kancrn.org/stream/

Notes:

Turbidity Test

1. Fill the test tube or glass with the sample. Continually shake or swirl the container to ensure the sediment stays suspended.

2. Place the test tube/glass over each circle, starting with E. Look down through the test tube/glass. If you cannot distinguish between the dark and light sections, move to the next circle (D).

3. The first circle where you can distinguish between the dark and light sections is your turbidity reading.

4. If you can distinguish between the dark and light sections in all of the circles, then you have mostly clear water with little turbidity.

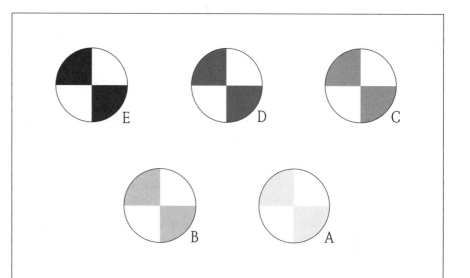

A–Little effect on aquatic plants and animals
B–Less light reaches plants, photosynthesis slows
C–Algae and zooplankton production drops
D–Aquatic insect production slows
E–Stressful for some fish due to lack of food production

Data Sheet for Turbidity Test

Water Sample 1:
    Letter:

    Interpretation:

Water Sample 2:
    Letter:

    Interpretation:

Water Sample 3:
    Letter:

    Interpretation:

Footprints On The Sand

Summary

Students simulate development of a beachfront community to explore the possible effects of development on water quality.

Objectives

Students will:

- distinguish between point and nonpoint source pollution.
- recognize that everyone contributes to and is responsible for a beach's water quality.
- identify causes of water pollution, the contaminants that are associated with pollutants, and the resulting health effects of pollution.
- identify Best Management Practices to reduce pollution.

Materials

Part I

- *Large sheets (6) of poster board, newsprint, or flipchart paper*
- *Copy of* **Beachfront Properties** *Teacher Copy Page*
- *Drawing pens and pencils*
- *Items from students' desks* (e.g., pencil, paper clip, book)

Part II

- *Copies of* **Water Contaminants Worksheet** *Student Copy Page* (1 per group)

Background

Maintaining good quality water in coastal areas is important because tourism, recreation industries, and

Grade Level:
Part I: 6–12
Part II: 9–12

Subject Areas:
Environmental Science, Chemistry, Health Science, Geography

Duration:
Preparation:
30 minutes

Activity:
Part I: 50 minutes

Part II: 30 minutes

Setting:
Classroom

Skills:
Interpreting, Modeling, Researching, Applying Solutions

Vocabulary:
nonpoint source pollution, point source pollution, stormwater runoff, contaminant, best management practices (BMPs)

fisheries all depend on it. Like any body of water, coastal areas are subject to pollution from many sources. The quality of coastal waters (like that of rivers and streams) largely reflects the land use practices on the surrounding land and the health of contributing watersheds. When watershed managers investigate land use practices that affect water quality they look for two general sources of pollutants: point and nonpoint source pollution.

Point source pollution involves pollutants that are discharged from, and can be traced back to, an identifiable point or source, such as a factory's discharge pipe or a sewage ditch. Nonpoint source pollution (NPS), on the other hand, cannot be traced to a single source, but is made up of many sources. Stormwater runoff and ground water transmit both kinds of pollution.

While point source pollution, once identified, can be directly mitigated, managing nonpoint source pollution can be much harder. Stormwater runoff can pick up and carry contaminants from large areas, thus concentrating and then distributing these pollutants into coastal waters. Because pollution comes from so many places and is of so many types, it is hard to manage and control. Polluted runoff can carry fertilizers, lawn chemicals, herbicides, road salts, oil and gasoline, untreated sewage and pet waste, among numerous other contaminants. It can even carry air pollutants like sulfur dioxides, nitrogen sulfides, and copper that precipitation has picked up from the atmosphere as it falls.

Polluted runoff can cause serious water quality and environmental problems for coastal areas. Common problems are beach closings, fishery and shellfish bed closures, habitat destruction, and toxic algal blooms. (See *Pfiesteria* sidebar)

While large point source events like coastal oil spills and oil tanker disasters often attract more attention, it is contributions of oil pollution from individual actions in cities across the world that account for major oil pollution of coastal waters. According to NASA (1995), the annual urban runoff from a city of five million people can contain as much oil and grease as a large tanker spill.

A National Research Council study showed that oil pollution of coastal waters stemmed primarily from nonpoint sources on land, especially from individuals dumping used motor oil. These "down the drain" sources accounted for more than all other sources of oil spills combined (363 million as compared to 343 million gallons), including major tanker spills, offshore drilling spills, natural seeps, etc.

The economic impact of water quality and environmental problems is enormous. Beaches remain the primary destination for vacationers. The NRDC (2001) rates the coastal tourist expenditures of sixteen states at $104 billion. Commercial fishing companies are also impacted when shellfish beds and fishing areas are closed due to bacterial contamination. For example, after the 1997 *Pfisteria* outbreak, the Maryland Sea

Beach closures pose serious health and economic impacts. Courtesy: Water Education Foundation

Pfiesteria piscicida and Coastal Pollution

"Dinoflagellates are a natural part of the marine environment. They are microscopic, free-swimming, single-celled organisms, usually classified as a type of alga. Most dinoflagellates are not toxic and obtain energy via photosynthesis. Others, including *Pfiesteria piscicida,* are more animal-like and acquire energy by consuming other organisms. *Pfiesteria piscicida* was discovered in 1988 and is believed to have a highly complex life cycle with 24 reported forms, a few of which can produce toxins" (Maryland Department of Natural Resources, 2000).

Although the dinoflagellate *Pfiesteria piscicida* has been around for thousands of years, it has only become a major problem, especially on the mid-Atlantic coast of the United States, since the 1980s. *Pfiesteria* can be a nontoxic predator of organisms such as bacteria, algae, or small animals, unless its environment is overly enriched with nutrients, particularly phosphorus. The excess nutrient load is a direct result of nonpoint source pollution from runoff carrying fertilizers and animal wastes downstream. *Pfiesteria piscicida* can multiply in the presence of increased phosphorus levels, and for poorly understood reasons, suddenly secrete harmful toxins. Other types of algal blooms produce brown and red tides, which also can be harmful to aquatic life.

Major fish kills are associated with *Pfiesteria piscicida* and *Pfiesteria*-like organisms. Ulcerative lesions and lethargy leave fish helpless to predators. Humans can also have severe reactions to *Pfiesteria piscicida.* Skin irritations, narcosis, reddening of the eyes, severe headaches, vomiting, impaired breathing, acute short-term memory loss, cognitive impairment, and blurred vision have all been reported by people who have contacted the organisms. Most of the symptoms disappear over time, but relapses have occurred after strenuous physical activity.

Grant Program estimated losses of $43 million in seafood revenue (National Oceanographic and Atmospheric Administration [NOAA], 2002).

In the U.S. in 2000, ocean, bay, lake, and freshwater beaches were closed or had swimming advisories for more than 11,000 days. Eighty-five percent of these closures were due to bacteria levels exceeding allowable water quality standards for human contact. In contrast, eight percent of the closures were due to known pollution events, such as sewage treatment plant failures or leaks. Six percent were precaution-ary closures due to a known pollution-causing event such as a rainfall overwhelming a treatment plant. Two percent of the closures were due to other causes, such as algal blooms, chemical spills, fish kills, strong waves, no lifeguards, or others (NRDC, 2001).

To address the issue of coastal pollution, the United States Environmental Protection Agency (USEPA) initiated the BEACH (Beaches Environment Assessment and Coastal Health Act) amendment to the Clean Water Act in October 2001. The BEACH amendment requires all states with coastal waters to adopt new or revised water quality standards by April 2004.

The protection of coastal waters from NPS pollution presents an enormous challenge because of the widespread and diverse nature of the problem. Land and water managers rely on methods called *Best Management Practices,* or BMPs, to describe measures designed to reduce or eliminate NPS pollution problems. A list of nonpoint source pollution sources and suggested BMPs can be found in the chart on page 93.

Best Management Practices for Reducing NPS Pollutants	
Source	**Best Management Practices**
Roads and Streets	• Dispose of paints, solvents, and petroleum products at approved disposal sites—not in storm drains or street gutters. • Fix automobile oil and fuel leaks. • Stop oil dumping on rural roads. • Use non-chemical de-icers (sand and ash) on roads, sidewalks, and driveways. • Construct a sediment catch basin to collect stormwater runoff. • Reduce road construction runoff by building terraces and catch basins, by using sediment fences, and by planting cover crops.
Construction	• Implement a sediment control plan. • Plant ground cover to reduce erosion. • Dispose of solvent, paint, and other wastes at approved disposal sites. • Build small, temporary dikes to slow and catch runoff. • Build berms to filter runoff before it enters water way. • Trap sediment with straw bale barriers.
Residential	• Use non-chemical de-icers (sand and ash) on residential driveways and sidewalks. • Read labels prior to using pesticides and fertilizers. • Clean up pet wastes and dispose of them in a landfill or compost pile. • Use non-chemical fertilizers (compost) on gardens. • Dispose of hazardous household waste, including motor oil, at approved sites. • Maintain vegetation buffer strips along waterways. • Channel runoff through vegetation to slow the water and to filter sediment and contaminants. • Maintain septic tanks. • Keep grass clippings and yard debris out of water ways.
Recreational Boating	• Dispose of human wastes at proper onshore sewage disposal facilities. • Minimize petroleum spills from bilges or when filling the tank. • Use care when applying hazardous maintenance products, such as marine paints and engine treatments. • Pack all trash and litter back to shore for disposal.

Procedure

Preparation

Draw the outline of the beachfront properties (shown on the *Beachfront Properties*) onto a poster board, sheet of newsprint, or flipchart paper. Color in the water with a blue marker. The number and type of land structures (treatment plant, parking lot, etc.) can vary from the chart. Improvise by leaving more areas free of development or adding structures that correspond with local conditions (e.g., fishing pier, marina, etc.). There is no maximum number of sheets required for this activity so add as many as needed.

Divide the board/sheet into equal parts giving each part some water-front and some land. Number the sheets in the upper left corner; this will allow the students to align their properties in the proper order to create an entire beachfront community. Make numbers readable but not prominent. The number of sections should correspond with the number of student groups. (For durability, the boards/sheets can be laminated.)

Warm Up

Ask how many students have ever attended a large gathering (concert, sporting event). What happens to all of the garbage left behind? Where did all of that garbage come from? Explain that each individual may have left just a little trash on the ground, but when it's combined with the trash of many others, the volume can be staggering. Water pollutants build to problematic size in a similar fashion—many small sources accumulate until the amount becomes a water quality issue.

Ask students if they have ever gone to a beach to go swimming; only to find it closed. Why was the beach closed? Ask students to brainstorm other reasons a beach might be closed for swimming (e.g., sharks, weather, pollution). List their ideas on the board.

The Activity
Part I
1. **Divide students into cooperative groups of three to five and tell them that they have inherited a piece of beachfront property and one million dollars. Have them list ways they could use the money to develop their beachfront.**
2. **Distribute a "piece" of the previously prepared beachfront property along with drawing implements to each group. Instruct students to draw how they plan to develop their land using their one million dollars.** Some of the properties already have structures on them, including a wastewater treatment plant, a pet-play park, a parking lot, and a boardwalk with restrooms. Some properties have no structures on them.
3. **When the drawings are complete, ask students to identify possible sources of water pollution on their property.** How have their development ideas affected water pollution?

4. **Once they have identified their sources of water pollution, have them look in the upper left-hand corner of their property for a number.** Explain that each piece of property is actually a part of a larger puzzle. Have students assemble their pieces numerically to create a stretch of beach.
5. **Each group should then describe how they developed their land and discuss any resulting sources of water pollution. Have them use items from their desks (e.g., book, pen, coin, paper clip, etc.) to represent these pollution sources.** As they describe the pollutants, the representative items should be placed directly on their property at the pollution source (e.g., paper clip on the parking lot to represent the runoff of petrochemicals from vehicles). Be prepared to mark other pollution sources that the students have overlooked.
6. **After the students have described their properties, have them move their pollution tokens into the water adjacent to their land.** Then mix them into one pile.
7. **Explain that coastal waters often have prevailing currents, and the current along this beach moves from the property with the pet play area toward the opposite end. Ask students how this affects their plans to use their land. Did they plan on having clean water adjacent to their property?** Move the representative objects in the direction of the

prevailing current until they are all in one pile at that end. This indicates the accumulated pollution from all their properties.
8. **Have the students try to reclaim their specific items. The items that are easily identifiable as theirs (e.g., keys, wallet, personalized items, unique pens or pencils) simulate point source pollution.** Items that are not so easy to identify as theirs (e.g., rubber bands, paper clips, coins, cans, common pens and pencils) simulate nonpoint source pollution: those pollutants that come from multiple, not readily identifiable sources.

Part II
1. **Hand out the *Water Contaminants Worksheet* to each group. Have students compare the pollutants from their property to those listed.**
2. **Instruct students to list the specific contaminants associated with their pollution sources.** For example, if their land was a source of pet wastes, they should list the contaminants contributed by this pollution—nutrients (phosphates and nitrates) and bacterial contaminants (coliform bacteria).
3. **If a group has a pollutant that is not listed on the *Water Contaminants,* have them conduct further research to find out about the specific contaminants and health impacts associated with that pollutant.**
4. **Have groups present their contaminants to the rest of the class.** What contaminants tended to

accumulate from different properties or from different pollution sources? What possible effects could the contaminants have on people using the beach? On the environment near the beach?

Wrap Up

After all the contaminants have been identified, discuss the activity with the students. Who is responsible for the beach pollution? Could a property on one end of a beach be affected by the actions of a property on the other end? How did those students who were down—current feel as they saw the accumulated pollution come by their property? How would this pollution affect their plans for using their property?

Ask students if they would swim in the water near their property considering the contaminants and their health effects? What steps could they take to reduce the contaminants and help ensure the water was safe?

As a follow-up, have each student write one paragraph detailing ways to reduce the pollution he or she contributed. (Share the *Major Sources of NPS Pollution and BMPs* from **Background.**)

Assessment

Have students:

- distinguish between point and nonpoint source pollution *(Part I).*
- identify causes of water pollution, the contaminants that are associated with pollutants, and the resulting effects on health

from pollution *(Part II).*

- recognize that everyone contributes to and is responsible for a beach's water quality (***Wrap Up***).
- identify *Best Management Practices* to reduce pollution (***Wrap Up***).

For further assessment, have students:

- design a beachfront community using *Best Management Practices* to minimize pollutants.

Extensions

Instead of a beachfront, have properties represent a lake system. Draw the properties encircling a lake; inlet and outlet streams can be added. Conduct the activity as described in the **Procedure.**

Another option is to conduct the activity without the drawing and have students form a river, lake, or beachfront. For example, one student represents a lake. A group of students encircle the "lake" to represent houses around the lake. Other students, standing in lines extending from the lake, are streams flowing to the lake. Students pass their pollution tokens downstream and into the lake until the person in the middle (the lake) holds all the items.

Complete the main activity using examples of real water users in the watershed where students live. Or assign roles (farmers, suburban dwellers, etc.) to students and have them develop their land accordingly. How would they manage their land

to protect water resources?

 Testing Kit Extension
Use Healthy Water, Healthy People Testing Kits to monitor your local beach for the contaminants listed in the activity. Ask a local water quality specialist to join the students in the field, or to visit your classroom prior to the field monitoring to discuss local water quality issues.

Resources

Maryland Department of Natural Resources. 2000. *What You Should Know About Pfiesteria piscicida.* Retrieved July 8, 2002 from the Web site: http://www.dnr.state.md.us/pfiesteria/facts.html

National Aeronautics and Space Administration (NASA). 1995. *Oceans in Peril; from Smithsonian's Ocean Planet.* Retrieved April 18, 2002 from the Web site: http://seawifs.gsfc.nasa.gov/OCEAN_PLANET/HTML/ocean_planet_oceans_in_peril.html

National Oceanographic and Atmospheric Administration (NOAA). 2002. *Polluted Runoff.* Retrieved April 2, 2002 from the Web site: http://www.ocrm.nos.noaa.gov/pcd/6217.html

Natural Resources Defense Council (NRDC). 2001. *Testing the Waters 2001: A Guide to Water Quality at Vacation Beaches.* Washington, D.C.: Natural Resources Defense Council.

The Watercourse. 1995. *Project WET Curriculum and Activity Guide.* Bozeman, MT: The Watercourse.

Beachfront Properties

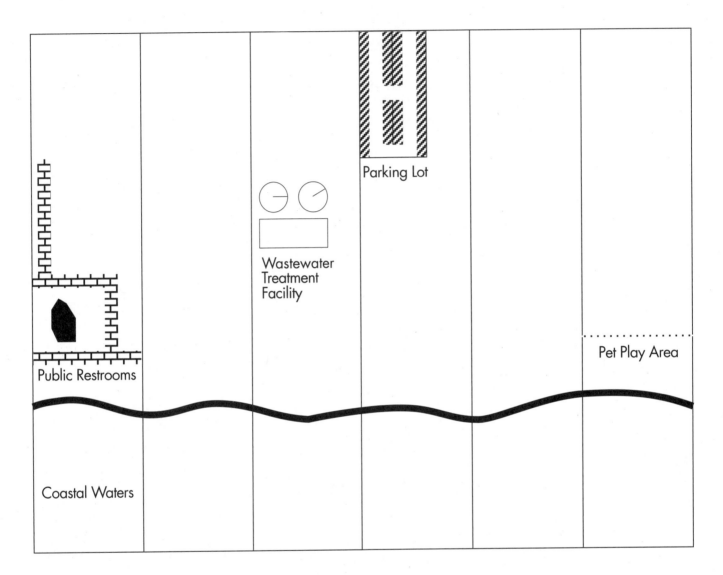

Water Contaminants Worksheet

Pollutant	Specific Contaminant	Possible Environmental and Human Health Impacts
Pet Waste	• Nitrate	Algal blooms–dissolved oxygen depletion
	• Phosphate	Algal blooms–dissolved oxygen depletion, *Pfiesteria*
	• Coliform bacteria	Human illness (watery diarrhea, fever, and dehydration)
Sediment	• Turbidity	Blocks sunlight, kills submerged vegetation
	• Phosphate	Algal blooms, dissolved oxygen depletion
	• Coliform bacteria	Human illness (watery diarrhea, fever, and dehydration)
Sewage	• Nitrate	Algal blooms, dissolved oxygen depletion
	• Coliform bacteria	Human illness (watery diarrhea, fever, and dehydration)
Litter	• Coliform bacteria	Human illness (watery diarrhea, fever, and dehydration)
Parking Lot Runoff	• Petroleum products (benzene, toluene, organic lead compounds.)	Human illness (fever, chills, vomiting)
Fertilizers	• Nitrate	Algal blooms, dissolved oxygen depletion
	• Phosphate	Algal blooms: human illness (respiratory illness)
Pesticides	• Chlorinated hydrocarbons, rhothane (DDD), lindane (HCH's), Atrazine	Human illness (headaches, dizziness, respiratory problems) Animal illness (muscle tremors, convulsions, tetanus)
Air Pollution	• Acid rain (sulfate aerosols, nitrogen oxides, copper, nickel, zinc)	Human illness (inherited heart defects, respiratory diseases, gastrointestinal disorders, skin and eye diseases, asthma, bronchitis, lung inflammation–asthma and emphysema)
	• Sulfur dioxide	Acidifies waters (lakes, rivers, streams)
Boat Pollution	• Sewage waste (bacteria)	Human illness (watery diarrhea, fever, dehydration, dizziness, muscle aches, vomiting)
	• Nutrient loading	*Pfiesteria piscicida* outbreaks; red-tide algal blooms
	• Litter	Animal strangulation (can block digestive system when ingested)
	• Oil spills	Human illness (cancer, sterility, brain dysfunction, fever, chills, ear discharge, vomiting)
Waterfowl	• Coliform bacteria	Human illness (watery diarrhea, fever, and dehydration)

WATER QUALITY
IS A VITAL
COMPONENT OF
HEALTHY COMMUNITIES
AND INDIVIDUALS

Looks Aren't Everything

Summary
Students study maps and clues from a hypothetical camping trip to determine how and why some of the campers became ill. They then investigate the role of water quality in human illness.

Objectives
Students will:
- calculate the amount of water they need in a day based on given weights.
- write a creative story or act out a skit describing their camping trip.
- analyze clues to determine which campers became ill and the cause of illness in campers at two of the campsites.
- use information from the activity to propose methods to ensure healthy water for future camping trips.

Materials
- Copies of **Camping Data Sheet** Student Copy Page (1 per group)
- Copies of **Situation Cards** Student Copy Page (1 card per group)
- Empty 1 liter, two liter, or gallon containers (enough for at least 15 liters total volume)

Background
Every day humans are inundated with different kinds of single-celled

Grade Level:
6–8

Subject Areas:
Biology, Mathematics, Language Arts, Environmental Science, Health

Duration:
Preparation: 10 minutes

Activity: two 50-minute periods

Setting:
Classroom

Skills:
Analyze, Evaluate, Calculate, Gather Information, Interpret, Present

Vocabulary:
waterborne disease, pathogen, unicellular, purification, protozoan, amoebae, Giardiasis, Cryptosporidium

organisms. Along with bacteria, which "are among the simplest, smallest, and most abundant organisms on earth" (Bartenhagen et al., 1995), we encounter protozoa, amoebae, and viruses. Some of these unicellular organisms are good, and some are harmful.

Helpful bacteria, for instance, are used to make substances that help our body function, or aid in digestion. Some antibiotics (which help us recover from sickness) come from bacteria, including penicillin, bacitracin, erythromycin, streptomycin, and tetracycline. Scientists are also studying bacteria's potential for cleaning up human-caused environmental contaminants such as pesticides, oil spills, and other hydrocarbons (Bartenhagen et al., 1995).

Unfortunately, some bacteria are pathogenic (producing toxins or causing diseases). Other single-celled organisms, like the protozoa *Giardia lamblia and Cryptosporidium parvum,* commonly contaminate fresh water sources in the United States, and often lead to human gastrointestinal illnesses.

Harmful unicellular organisms can be spread through water, soil, air, human contact, or fluid exchanges, (i.e., blood or saliva). Each type of organism is unique in how long it can survive in certain environments and in which environments it spreads. For example, *Giardia* is a common waterborne pathogen that forms a cyst and can lie dormant for long periods.

Regardless of how single-celled organisms are spread, their presence in water is undetectable by our unaided eyes, nose, or mouth. Tests must be performed in order to confirm or deny the presence of these harmful organisms in water. The United States Environmental Protection Agency (USEPA) has established a limit of zero bacteria in drinking water at municipal water systems, and therefore water treatment plants must test for them regularly.

However, when people are away from municipally treated water sources (camping or backpacking for instance), they must sometimes rely on natural sources for their drinking water. If such water is consumed without special precautions, unicellular organisms living in the water can cause serious illness. Two of the most common waterborne diseases that a camper might encounter are discussed below.

The best way to avoid these harmful organisms when hiking or camping is to carry your own drinking water. Because of water's sheer weight and the length of time spent outdoors, it is sometimes impossible to carry all that is needed and natural sources of water must be used. In that case, to lessen the chance of ingesting a harmful organism, it is best to collect drinking water from a flowing source rather than a stagnant one. If possible, avoid water that is odiferous, excessively turbid (suspended sediments), or dark in appearance.

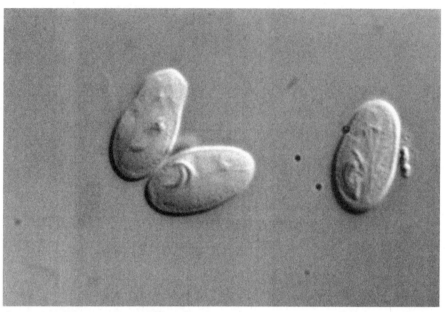

Giardia lamblia *magnified 1000x.* Giardia *is a common pathogen found in surface waters near campsites. American Water Works Association*

Waterborne Diseases
(American Water Works Association [AWWA], 1999)

Giardiasis
(*Giardia lamblia*)–commonly called Beaver Fever
- Transmission: contact with water, food, animals, or people that have had contact with fecal matter containing *Giardia*. Commonly carried by wildlife in ponds and streams. Typically takes 10-14 days for symptoms to appear.
- Symptoms: Diarrhea, weight loss, flatulence, cramps, belching, distention, anorexia, vomiting, fatigue, mucus in the stool, and bloody, foul-smelling stool.
- Treatment: quinacrine, metron idazole, furizolidone, and paramomycin.

Travelers' diarrhea
(*Escherichia coli*)
- Transmission: contact with water, food, animals, or people that have had contact with fecal matter containing *E. coli*. Animal fecal matter near a water source is the most common source of contamination. Can cause symptoms overnight.
- Symptoms: Watery diarrhea, fever, and dehydration.
- Treatment: Antibiotics.

Another option is to sufficiently purify water from natural sources to kill or remove harmful organisms.

Below are descriptions of the most common treatment methods.

Filtering:
To remove protozoa, filters should ensure at least **one** of the following levels of filtration (Center for Disease Control [CDC], 1999):
- Reverse osmosis
- Absolute pore size of 1 micron or smaller (NOT *Nominal* pore size)
- Tested and certified by NSF International Standard 53 for cyst removal
- Tested and certified by NSF International Standard 53 for cyst reduction

Boiling:
The EPA recommends vigorously boiling water for five minutes to kill harmful pathogens and to account for potential boiling point differences due to altitude (EPA, 2001).

Chemical Additions:
Both chlorine and iodine can be used to chemically treat water for drinking, though some pathogens (i.e., *Cryptosporidium*) are unaffected by chemical treatment and must be removed by filtering. It is easiest to use prepared chlorine or iodine tablets to treat water. These are often found in drug or sporting goods stores.

A backcountry campsite in Yellowstone National Park. Courtesy: Yellowstone National Park, NPS.

Procedure

Warm Up

Ask students what can happen if a person drinks untreated water from a stream, pond, or lake. Why don't most people drink water directly from these water sources? Inquire if they have ever been camping or hiking, and if so, how did they get water? Did they carry it with them? Did they treat it? Did they drink from a creek or spring? Have students share stories of people they know who became ill from drinking contaminated water, especially when camping or hiking.

Remind students that the water they drink from the school water fountain is treated and tested to ensure that it does not contain pathogens. When camping or hiking however, the best way to avoid contact with water-borne pathogens is to carry clean water or treat it along the way.

Tell students they are preparing for an overnight camping trip. They need to take enough water for the next three days—one day to hike in for five miles, one day at the campsite, and one day to hike out. In small groups, have students brainstorm what they will need water for during their trip (the hike in, drinking, cooking, washing, brushing teeth, washing dishes, hike out, etc.). To give them an idea of how much water they will need, have students complete Part I of the ***Camping Data Sheet*** and then share their results with the class.

How much will they have to carry as a group?

Before the groups decide if they want to carry that much water into their campsite, pre-measure the total liters of water needed by a group using the empty liter and gallon containers. It's important that each student get a chance to *feel* the weight of the water needed. *(Option: if there are enough containers, have each group measure the volume of water that they need.)*

The Activity

1. **Now the virtual camping trip begins! Instruct students to review the Campsite Map on their *Camping Data Sheet*.**
2. **Provide each group with a different *Situation Card*. Have groups study their card and plot their campsite location on the map.** Have them mark trails to the water sources nearest their campsites.
3. **Have them determine from the clues on their *Situation Card* how they obtained their water— did they carry it with them or retrieve it from a nearby pond or stream?** Explain that some water can harbor harmful bacteria and protozoa that can cause illness if the water is consumed without treating or filtering. According to their ***Situation Card***, the groups discover that some of them elected to drink water from local sources.
4. **To summarize their scenario, have students create a story or act out a skit of their experiences during the hypothetical camping trip.** Each point on their cards should be woven into their stories. Encourage creativity beyond these basic points. Have student share their stories and skits with the class.
5. **After the final group has presented, read the following situation:** *The school nurse has discovered a problem resulting from the camping trip. Severe illnesses have overcome two of the camping groups. One group of campers became ill the day they returned from the trip and the primary symptoms were diarrhea and fever. A second group did not display symptoms of illness until almost two weeks after the trip, and their primary symptoms were diarrhea, cramps, vomiting, and fatigue.*
6. **Challenge each group to predict which two groups became ill, the causes and types of illness contracted by these two groups, and possible reasons the other campers stayed healthy.** Have each group share their predictions with the class.

Wrap Up

As a class, discuss the victims and possible causes of the illnesses (Beaver Pond: *Giardia*–animals in the pond; Rattlesnake Mesa: *E. coli*– animal fecal matter around the water hole). Read the disease description from the ***Background*** if needed to give students an overview of the two illnesses.

Have students predict why the other groups did not become ill (e.g., carried own water or sports drinks

in, used spring water, boiled the water for hot chocolate before drinking). Did they notice anything similar about the water used by the campers? Did some groups boil their water and others not?

Have each student write a brief paragraph describing the ways they used water on their virtual camping trip and where the water came from. Include how they would ensure that they would have clean, healthy water to drink if they were to camp there again.

Assessment
Have students:
- investigate the challenges of obtaining water for a virtual camping trip (***Warm Up, Step 3***).
- calculate the amount of water they need in a day based on given weights (***Warm Up***).
- write a creative story or act out a skit describing their camping trip (***The Activity–Step 4***).
- analyze clues to determine which campers became ill and the cause of illness in campers at two of the campsites (***The Activity–Steps 5,6***).
- use information from the activity to propose methods to ensure healthy water for future camping trips (***Wrap Up***).

Extensions
Have student groups research various water filters and purifying treatments. Have them consider bringing equipment from home,

researching on the Internet, or inviting a local outdoor product expert to demonstrate several camping filters and treatments. Assign each group one type of filter or treatment to research and compile their results in a matrix displaying the benefits and limitations of the different methods. Matrix may include cost, portability, effectiveness against the different pathogens, ease of use, volume of water effectively treated by each method, etc.

Have students research the implications of living where water treatments are unavailable.

Have students develop a public education program on the health effects of drinking unsafe water. They should include possible emergency measures to take at home in the event of a natural disaster or temporary breakdown of the community's water purification system. See http://www.epa.gov/ogwdw000/faq/emerg.html

 Testing Kit Extensions
Have students use Healthy Water, Healthy People Testing Kits to test local waters for possible bacterial contamination such as total coliforms.

Using Healthy Water, Healthy People testing kits, have students evaluate the water purification effectiveness of boiling water versus using iodine treatment by comparing the levels of contamination in water treated by each method.

Have students design and conduct an experiment to determine how long water must be boiled to completely rid it of bacteria. Healthy Water, Healthy People testing kits can be used to measure student results.

Resources
American Water Works Association (AWWA). 1999. *Manual of Water Supply Practices: Waterborne pathogens.* Denver, CO: American Water Works Association.

Bartenhagen, K., M. Turner, and D. Osmond. 1995. *Bacteria.* Retrieved on January 15, 2002 from the North Carolina State University WATERSHEDDS: Water, Soil and Hydro Environmental Decision Support System Web site: http://h2osparc.wq.ncsu.edu/info/bacteria.html

Center for Disease Control (CDC). 1999. *Preventing Cryptosporidiosis: A guide to water filters and bottled water.* Retrieved on January 18, 2002, from the Web site: http://www.cdc.gov/ncidod/dpd/parasites/cryptosporidiosis/factsht_crypto_prevent_water.htm

Environmental Protection Agency (EPA). 2001. *Emergency Disinfection of Drinking Water.* Retrieved on January 17, 2002, from the Web site: http://www.epa.gov/ogwdw000/faq/emerg.html

Notes:

Camping Data Sheet

Part I

An active person needs two thirds of an ounce of water per day per pound of body weight. Calculate the amount of water that a fictitious group of campers will need per day by completing the chart below using the following equations:

1. (Body weight in pounds) X (0.67 oz.) = __oz. of water per day
2. Convert to liters: 1 fluid ounce = 30 ml or 0.03 liters; therefore
(__ oz. of water per day) X (0.03 liters) = __ liters of water per day

Body Weight x (0.67) = ounces per day x 0.03 liters = liters per day

Camper 1   100 lbs. x (0.67) = _____ ounces p/day x (0.03) = _____ liters p/day

Camper 2   118 lbs. x (0.67) = _____ ounces p/day x (0.03) = _____ liters p/day

Camper 3   120 lbs. x (0.67) = _____ ounces p/day x (0.03) = _____ liters p/day

Camper 4   107 lbs. x (0.67) = _____ ounces p/day x (0.03) = _____ liters p/day

Camper 5   115 lbs. x (0.67) = _____ ounces p/day x (0.03) = _____ liters p/day

Total liters per day needed for entire group: _____ liters per day

Campsite Map Student Copy Page

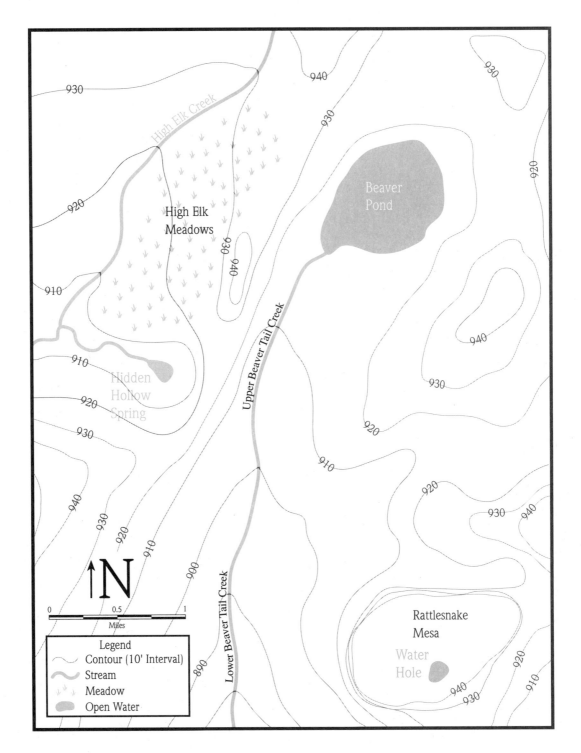

Situation Cards

Camping Location: Rattlesnake Mesa
Shelter Type: tent; 2 members slept under the stars
Water Source: water hole
Evening Meal: spaghetti, brownies, lemonade from powdered mix
• Saw a rattlesnake and other wildlife visit the water hole
• Great views of the entire area, cold clear night under the stars, good star gazing

Camping Location: Beaver Pond
Shelter Type: lean-to
Water Source: pond
Evening Meal: stew, rolls, peanut butter cookies, water
• Saw wildlife in meadow, saw beavers in the pond
• One member fell into the beaver pond while crossing the dam

Camping Location: High Elk Creek
Shelter Type: tents
Water Source: creek
Evening Meal: hot dogs, chili, s'mores, bottled sports drink which they carried in
• Saw an eagle; viewed wildlife in the meadow
• briefly rained on group, fast-moving creek was soothing for sleep
• Flaming marshmallow caught tent on fire briefly

Camping Location: Beaver Tail Stream, lower
Shelter Type: lean-to
Water Source: carried in all the water they needed
Evening Meal: burritos, tortilla chips, apple crisp, hot chocolate
• Heard unidentified thrashing in the underbrush near camp
• Caught fish in the swift stream; several members fell into the water because of slippery rocks
• Found plenty of wood to build lean-to, drank lots of hot chocolate to keep warm

Camping Location: Hidden Hollow
Shelter Type: tents
Water Source: spring
Evening Meal: pita pocket pizzas, cake, iced tea
• Enjoyed the fresh spring water, bubbling up from the ground
• Saw wildlife and lots of birds
• Burned dinner; tent fell down when group member tripped over its strings

Camping Location: Beaver Tail Stream, upper
Shelter Type: lean-to
Water Source: carried in all the water they needed
Evening Meal: chicken, baked potatoes, chocolate chip cookies, packaged fruit drinks
• Went wading in the stream looking for unique rocks, lean-to fell down during the night
• Saw deer along the stream, leftover food was stolen by raccoons
• Saw a fox drinking from the stream early the next morning

Setting the Standards

Summary
Students simulate the process used by the United States Environmental Protection Agency to determine drinking water standards.

Objectives
Students will:
- investigate the need for and value of water quality standards.
- simulate the EPA process of risk assessment and drinking water standard setting.
- compare and contrast risk characterizations from each group.

Materials
- Copy of **Drinking Water Standards** Student Copy Page (1 per student)
- Copies of **Risk Assessment Worksheet** Student Copy Pages (1 per group)
- Copies of **What is the Risk?** Student Copy Pages (1 per group)

Background
Though we all drink water every day, most people rarely think about drinking water standards—what they are, why we need them, where they come from.

The Safe Drinking Water Act (SDWA) of 1974 allows the United States Environmental Protection Agency (EPA) to determine drinking

Grade Level:
9–12

Subject Areas:
Environmental Science, History, Political Science, General Science, Geology, Chemistry

Duration:
Preparation:
1 hour

Activity:
Two 50-minute periods

Setting:
Classroom

Skills:
Synthesize, Analyze, Interpret, Simulate

Vocabulary:
micrograms per liter (μg/l), parts per billion (ppb), methyl tertiary butyl ether (MTBE), primary drinking water standard, secondary drinking water standard, oxygenate, impermeable

water standards, which regulate the level of contaminants deemed unsafe or unacceptable for human consumption. These standards "protect drinking water quality by limiting the levels of specific contaminants that can adversely affect public health and are known or anticipated to occur in water" (EPA, 2000, p.1). The EPA enlists assistance from various organizations to synthesize all of the available research, studies, technology, and cost considerations to determine these drinking water standards. The process involves not only scientific data, but also decisions from political, economic, and social perspectives. There are two types of drinking water standards: National Primary Drinking Water Standards and National Secondary Drinking Water Standards.

Primary drinking water standards regulate contaminants that are hazardous to human health. A maximum contaminant level (MCL) determines the allowable concentration of each contaminant in drinking water. Public water treatment facilities are legally obligated to treat water to keep the concentration of contaminants less than or equal to the MCL. For example, the MCL for mercury is 0.002 mg/l, or part per million (ppm). Therefore, water treatment facilities can allow 0.002 mg/l or less of mercury in drinking water.

Secondary drinking water standards regulate contaminants that do not pose any health risks but are either cosmetically or aesthetically unpleasant to consumers. Cosmetic effects include water that can cause discoloration of skin or teeth. Aesthetic effects include unpleasant color, taste, or odor. Though these standards are not legally enforceable, they do serve as guidelines for public water facilities to follow. For example, the secondary drinking water standard for iron is 0.3 mg/l. Although iron is an essential mineral for humans, at high levels it stains toilets, sinks, and bathtubs, so consumers prefer lower levels of iron.

Even though the EPA has the authority to establish drinking water standards, they do not work alone. Scientists provide data from studies on the health effects of contaminants, where they occur, how they move through the environment, and how to remove them from water.

The EPA is also advised by the National Drinking Water Advisory Council (NDWAC), a fifteen-member panel of representatives from private organizations, state and local water supply or water hygiene agencies, and the general public.

Scientists, the NDWAC, and other interested parties follow an established Risk Assessment Process (see illustration page 112) to determine water quality standards. This process begins with identifying a hazard that is "(1) not already regulated by the SDWA; (2) may have adverse health effects; (3) is known or anticipated to occur in public water systems; and (4) may require regulations under the SDWA" (EPA, 2000, p.3). Hazards that meet these criteria are placed on the Contaminant Candidate List (CCL). Scientists then conduct an exposure assessment and a dose-response assessment on the listed contaminant.

An exposure assessment determines where the contaminant occurs, how it is transported, and where it is stored in the environment. From this information, the degree and duration of human exposure can be determined.

In a dose-response assessment, scientists study the health risks associated with different doses of the contaminant. For example, they might determine the health risk associated with a low dose or very small amount of the contaminant compared with a large dose of the contaminant. They consider how both healthy populations and vulnerable populations (the elderly, infants, immuno-compromised) are affected by the contaminant. The short-term and long-term health effects of the contaminant are also studied. This information is interpreted in a process called risk characterization.

Finally, the costs of implementing a standard are considered. Because water treatment facilities must often purchase costly equipment in order to meet a new standard, they are given three years to comply. Usually the whole process takes about seven years (EPA, 2001).

As an example, consider one contaminant, methyl tertiary butyl ether (MTBE) now on the Drinking Water

Maintaining healthy drinking water is the goal of setting water quality standards. Courtesy: US Environmental Protection Agency

Contaminant Candidate List (CCL). About 1979, MTBE replaced lead as an octane-enhancer in gasoline. Several years later, it was discovered that MTBE improved the combustion of gas, which also reduced carbon monoxide emissions. In about 1992, the EPA mandated the use of MTBE as an oxygenate to reduce carbon monoxide emissions in certain cities that exceeded carbon monoxide limits established by the Clean Air Act Amendments of 1990. (Incidentally, ethanol is also an oxygenate, but posed logistical complications that made MTBE the preferred oxygenate.)

Upon inception of the 1992 EPA mandate, residents near gas stations along with consumers began to complain of watery eyes, headaches, and nausea. They also complained of an unpleasant taste and odor in their drinking water. Upon investigation, it was determined that underground gasoline storage tanks, containing MTBE, were leaking fuel into the ground water.

Because MTBE dissolves easily in water, it tends to stay in ground water rather than adsorbing into soils and organic matter as gasoline does. So far the EPA has implemented a secondary drinking water standard of 20 to 40 micrograms per liter (μg/l) or parts per billion (ppb) of MTBE. Exposure assessments and dose-response assessments are currently underway. Upon completion of the MTBE studies, they will assess the need for a primary drinking water standard, and if needed, an MCL will be determined.

Procedure

Warm Up

a. Fill a clear plastic cup with tap water from a school water fountain. Tell the students that the cup contains water from an unknown source. Ask them whether they would drink it or not and to explain their decision. How many students said they would *not* drink the water?

b. Explain to the students that the water came from the school water fountain—now would they drink it? (Hopefully they will, knowing the source.) What made them trust the water from the water fountain over the water from an unknown source?

c. Hand out the **Drinking Water Standards** Student Copy Page and explain that in the U.S., the EPA requires all drinking water treatment plants to meet these standards to protect human health. Describe the differences between a primary and secondary water quality standard, along with the term maximum contaminant level, or MCL. Ask students how they think the EPA may arrive at these standards?

d. Draw the EPA Risk Assessment Process flow chart on the board or place a copy on an overhead projector. Lead the students through the following steps, which introduce the EPA process for setting drinking water standards. *Note: You may ask stu - dents to conduct further research on a specific contaminant of their choosing in order to answer the following questions.*

Step 1: Hazard Identification
Have students brainstorm contaminants that might be found in drinking water, and list them on the board. Examples include sediment, bacteria, nitrate, oil, heavy metals (mercury, arsenic), etc.

Step 2: Exposure Assessment
Ask students individually or in small groups to choose a contaminant and speculate where this contaminant might come from and how it might be transported through the environment. (Is it found in water, soil, or the air? How did it get in the water?)

Explain that the municipal drinking water treatment process can remove some contaminants and not others. Pollutants like heavy metals and oils are more difficult to remove. Also, well water is often untreated and can contain nitrate, metals, or other pollutants.

Step 3: Dose-Response Assessment
The presence of contaminants in water doesn't necessarily mean people will become ill. Much depends on the specific contaminant and the strength of the affected individual's immune system. Generally, infants, the elderly, and immuno-compromised individuals are more likely to fall ill from contaminated water. Even people with strong immune systems can become sick when exposed to small

doses of a contaminant over a long period of time. Ask students to speculate how harmful their chosen contaminant may be to human health.

Step 4: Risk Characterization
Have students refer to the Drinking Water Standards Chart on pages 118-120 to determine the maximum contaminant level (MCL) of their chosen contaminant. According to the EPA Drinking Water Standards, is their contaminant a primary or secondary contaminant?

The Activity

Tell students that they will be simulating the process that the EPA follows in establishing drinking water standards.

1. Inform students that the Safe Drinking Water Act (SDWA) of 1974 allows the EPA to determine drinking water standards. This lengthy process involves input from concerned citizens; local, state, federal, and tribal agencies; private organizations; environmental groups; toxicologists; epidemiologists; geologists; and representatives from water utilities.

2. Form groups of 3-5 students, and read the following scenario aloud: *"The year is 1992. Local residents have recently been complaining of watery eyes, headaches, and nausea when pumping gas or when in proximity to gas stations. Residents have also noticed their drinking water, which is primarily from wells, has developed a strange taste and odor. Similar complaints have occurred in*

large cities across the nation.

Scientists speculate the health symptoms are related to a gasoline additive. The Environmental Protection Agency (EPA) recently implemented a mandate in several cities across the United States to improve the combustion of gasoline as a way of reducing carbon monoxide emissions. To satisfy the mandate, gasoline was mixed with methyl tertiary butyl ether (MTBE) a chemical that increases fuel oxygen, which in turn increases combustibility. Fumes from the additive, MTBE, may be causing the symptoms reported by the local residents in people pumping gas. In addition, a local gas station recently reported a leaking underground storage tank (LUST) on their property, which is adjacent to the municipal water treatment plant. This could account for the strange new taste and odor of the local drinking water."

3. Distribute the *Risk Assessment Worksheet and What Is the Risk?* to each group. As a class, discuss and complete question one on the *Risk Assessment Worksheet.* Answers should include health symptoms, such as watery eyes, headaches, and nausea, as well as taste and odor complaints about drinking water.

4. Ask groups to complete the rest of the worksheet on their own using information gleaned from *What Is the Risk?*

5. Have a representative from each group report their conclusion to the rest of the class. Students should be able to propose

an action, and defend that action for the organizations listed in problem four (risk characterization) of the worksheet. To facilitate a discussion, you may wish to write the organizations and the groups' recommendations on the board.

6. Discuss the various actions proposed for each organization.

Wrap Up

Discuss with students why the proposed actions might vary between groups. Is any answer right or wrong? What additional information would groups request? How might new information change their decisions? Inform students that the EPA is currently investigating MTBE and, in the meantime, has established a secondary drinking water standard between 20 and 40 micrograms per liter (μg/l) or parts per billion (ppb). The EPA is currently collecting data on the oral dose-response assessment of MTBE, and will likely establish a primary drinking water standard based on the dose-response results.

Assessment
Have students:
- investigate the need for and value of water quality standards (**Warm Up**).
- simulate the EPA process of risk assessment and drinking water standard setting (**Steps 2–6**).
- compare and contrast risk characterizations from each group (**Wrap Up**).

Extensions
Research the Web sites in the

resource section below to determine if the EPA has issued its final recommendations on MTBE standards. How do these standards compare with the student recommendations?

Have students write an essay in favor of establishing a standard for MTBE from the point of view of one of the following interest groups (ensure that they consider political, economic, social, as well as scientific ramifications):
Drinking water treatment plant manager; mayor of city government that pays for drinking water treatment; gas station owner; air quality monitoring group spokesperson; elderly person with breathing difficulty; parent of a newborn; nearby resident.

Use a Ground Water Flow Model to simulate the movement of MTBE through the soil and into ground water. Have students write a description of the process by which contaminants can enter the ground and end up in drinking water.
Before and/or after conducting the activity, ask student groups to formulate their own risk assessment process. Compare this process with the EPA risk assessment process.

Testing Kit Extensions

Obtain the current drinking water standards from the EPA Web site: http://www.epa.gov/safewater/mcl.html. Perform tests on well water, tap water, pond water or other water sources to see which ones meet the current drinking water standards.

Resources

Environmental Protection Agency (EPA). 1997. *Drinking Water Advisory: Consumer Acceptability Advice and Health Effects Analysis on Methyl Tertiary-Butyl Ether (MTBE)*. Retrieved on February 12, 2002, from the Web site: http://www.epa.gov/waterscience/drinking/mtbe.pdf

Environmental Protection Agency (EPA). 2000. *Setting Standards for Safe Drinking Water*. Retrieved on February 11, 2002, from the Web site: http://www.epa.gov/safewater/standard/setting.html

Environmental Protection Agency (EPA). 2001. *Drinking Water Contaminant Candidate List*. Retrieved on February 18, 2002, from the Web site: http://www.epa.gov/safewater/ccl/cclfs.html

Environmental Protection Agency (EPA). 2001. *MTBE/Oxygenates*. Retrieved on February 15, 2002, from the Web site: http://cfpub.epa.gov/ncea/cfm/oxygenates.cfm

Environmental Protection Agency (EPA). 2002. *Current Drinking Water Standards*. Retrieved on February 22, 2002, from the Web site: http://www.epa.gov/safewater/mcl.html

Zogorski, J, M. Moran, and P. Hamilton. 2002. USGS examines MTBE. *Water Technology,* 25 (3), 26-30.

Notes:

Risk Assessment Worksheet

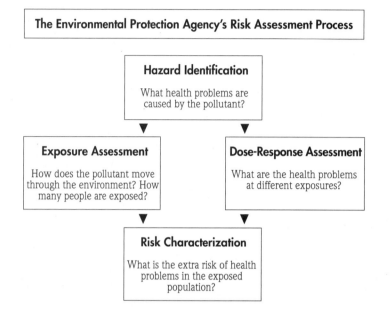

The Environmental Protection Agency's Risk Assessment Process

Hazard Identification

What health problems are caused by the pollutant?

Exposure Assessment

How does the pollutant move through the environment? How many people are exposed?

Dose-Response Assessment

What are the health problems at different exposures?

Risk Characterization

What is the extra risk of health problems in the exposed population?

1. **Hazard Identification:** After hearing the scenario, why might you think MTBE is a potential hazard? List several reasons.

2. **Exposure Assessment:** Briefly summarize the Exposure Assessment data from *What Is the Risk?*
   Geologic Survey:

   Ecological Site Survey:

   Properties of MTBE:

   Surface Soil Sampling:

   Surface Water Sampling:

   Ground Water Sampling:

3. **Dose-Response Assessment:** Briefly summarize the Dose-Response data from *What Is the Risk?*
Cancer effects:

Non-cancer toxicity:

Taste and odor:

4. **Risk Characterization:** Hazard Identification: Would you consider MTBE a hazard? Why or why not?

Exposure Assessment: Will MTBE move through the environment? Will humans be exposed to MTBE if it does move through the environment?

Dose-Response Assessment: Are there health problems associated with MTBE?

Is there other information not provided that you feel is relevant to characterizing this risk? Based on the data provided, as well as any other information or concerns you may have, what course of action would you suggest for the following organizations?

Wildlife manager of this region:

Municipal water treatment plant manager:

Environmental Protection Agency board for establishing primary drinking water standards:

Environmental Protection Agency board for establishing secondary drinking water standards:

Local residents:

What Is the Risk? Worksheet

The municipal water supply in your city comes from both surface water and ground water (water that is naturally stored underground). Water is pulled directly from the river to the water treatment plant. One well (Town Well 1) is used to pump the water from the ground to the water treatment plant. The well is 45 feet deep. The river, the well and the water treatment plant are located just outside of town. The land surrounding the treatment plant belongs to the state and is unoccupied.

The closest commercial property is a gas station. The pump manager at the gas station noticed profits had been slowly decreasing for several months. She was puzzled since the population was growing in the area and business had been steadily increasing. After serious investigation, she found that they were not selling as much gas as they were purchasing. From this, she concluded that their storage tank was leaking. The storage tank is 35 feet below ground. She reported this information to the Department of Environmental Quality (DEQ).

The DEQ suspected a connection between this information and the increased number of calls from local residents complaining about the strange taste and odor of their drinking water. When the DEQ ordered that the drinking water be tested small amounts of MTBE (methyl tertiary butyl ether) were found. Then, a full-scale risk assessment began.

Information supplied by local toxicologists, geologists, concerned citizens, and others has been given to you. You must analyze the information and answer the associated questions on the *Risk Assessment Worksheet*.

I. Exposure Assessment

1. **Geological Survey:** An investigation of the local geology indicates that glacial till underlies most of the site, grading to a silty composition on the banks of the river. There is an impermeable clay layer about 15 feet below ground level that prevents water from passing through this layer. The water table, or the upper surface of the ground water, begins at 30 feet below ground level.

    Local ground water modeling indicates that ground water follows the surface topography, flowing downhill from the north side of the site toward the river. Water moves quickly and easily through glacial till because of its large pore spaces (spaces in-between rocks).

    Mark with an arrow on the picture the direction in which ground water moves.

    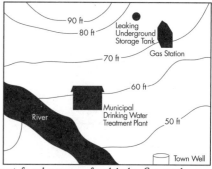

2. **Ecological Site Survey:** The area surrounding the ground water treatment plant and the gas station is covered with native vegetation—grasses, wildflowers, and trees. No threatened or endangered species live here. In fact, very few animals inhabit the area.

    The river teems with insects and invertebrates, which serve as an excellent food source for birds. Several trout were spotted from the banks, indicating the water temperature is cool. A rapid bioassessment of macroinvertebrate populations determined the river is in excellent condition.

    What findings can you determine about the ecological health of the area?

3. **Properties of MTBE:**

    Chemical formula: $C_5H_{12}O$

    Boiling point: 55.2 °C

    Solubility: 4.8 g/100 g of water This is highly soluble, meaning it dissolves easily in water and will move at the same rate as the ground water. Because of it high solubility, it also tends to be difficult to remove from water

    Volatility: extremely volatile, it evaporates easily

    Fat solubility: very low, does not accumulate in soils, sediments, and organisms—tends to stay in water or air

    Eco-toxicology value: 34,000 µg/l (ppb)—below which there is no expected adverse health effect to any animal species

    Human Taste and Odor Threshold: 40 µg/l (ppb) below which humans cannot detect the presence of MTBE

    Which of these properties might be helpful when assessing risks associated with MTBE? Does any of this information help you determine where in the environment you should focus your investigation? Based on this information, should you look at ground water samples, soil samples, surface water samples?

    If humans cannot detect MTBE below about 40 µg/l (ppb), how might this information be useful in establishing a secondary standard for taste and odor?

4. **Surface Soil Sampling:** Soil samples were collected from four sites between the gas station and the municipal water treatment plant. The top five inches of soil were collected from each site, and analyzed for MTBE contamination. The map indicates the site locations.

    Based on this data, should the DEQ be concerned about the soil around the gas station or the ground water treatment plant? Should they be concerned about wildlife and humans that work around these sites?

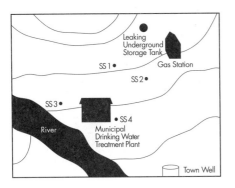

MTBE concentration in Surface Soil Samples (µg/l or ppb)			
SS 1	SS 2	SS 3	SS 4
None detected	None detected	None detected	None detected

5. **Surface Water Sampling**: Water samples were collected from four sites within the river. Three were taken from near the banks, and one from the middle of the river. The samples from the banks were collected from the surface, while the sample from the middle of the river was collected between the surface and the riverbed. Each sample was analyzed for MTBE contamination. The map indicates the site locations.

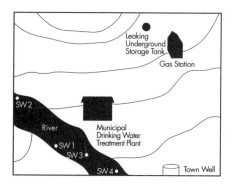

MTBE Concentration in Surface Water Samples (µg/l or ppb)			
SW 1	SW 2	SW 3	SW 4
None detected	None detected	None detected	None detected

Based on this data, should the water treatment plant be concerned about using the river as a source for drinking water? Should you be concerned about aquatic organisms and wildlife that rely on the river? Why might contaminated ground water from this site not contaminate the river? Hint: think about the geology.

6. **Ground Water Sampling**: Four temporary sampling wells were dug to the depth of the Town Well 1, which is 45 feet. All of these wells were dug down gradient (which in water terms means downhill) from the leaking underground storage tank to determine if it was the source of contamination in the drinking water. Each sample was analyzed for MTBE contamination. The map indicates the well locations.

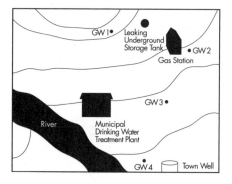

MTBE Concentration in Ground Water Wells (µg/l or ppb)			
GW 1	GW 2	GW 3	GW 4
None detected	1,500	800	75

Do you think the amount of MTBE in ground water Wells 3 and 4 will be increasing or decreasing in the future? Should the municipal water treatment plant be concerned about their ground water source?

II. Dose-Response Assessment

Dose-response studies try to link the quantity of a contaminant to its associated health risks. Because dose-response data are not yet available for oral ingestion of MTBE, the EPA has converted inhalation data to oral ingestion data. This process potentially introduces a high degree of error. Though scientists are studying the direct effects of MTBE contaminated drinking water; results are not available at this time. Therefore, base decisions on current data.

1. **Cancer effects:** The current data suggest that high doses (8,000,000 ppb) of MTBE have a potential to cause cancer in humans. Data is currently unavailable on the potential for low doses of MTBE to cause cancer.

2. **Non-cancer toxicity:** Current data suggest that high doses (8,000,000 ppb) of MTBE pose a health hazard to humans. Data is currently unavailable on the potential for low doses of MTBE to pose a health hazard.

3. **Taste and odor:** While human sensitivity to taste and odor varies greatly, current data suggest that most humans cannot smell or taste MTBE in water at concentrations between 20 and 40 micrograms per liter (μg/l) or parts per billion (ppb).

Student Copy Page

Drinking Water Standards I

Contaminant (mg/L)[2]	MCLG[1]	MCL or TT[1] (mg/L)[2] MCL	Potential health effects from exposure above the water	Common sources of contaminant in drinking water
Total Coliforms (including fecal coliform and *E. coli*)	Zero	5.0%[4]	Not a health threat in itself; it is used to indicate whether other potentially harmful bacteria may be present[5].	Coliforms are naturally present in the environment; as well as feces; fecal coliforms and *E. coli* only come from human and animal fecal waste.
Turbidity	N/a	TT[3]	Turbidity is a measure of the cloudiness of water. It is used to indicate water quality and filtration effectiveness (e.g. whether disease-causing organisms are present). Higher turbidity levels are often associated with higher levels of disease-causing microorganisms such as viruses, parasites, and some bacteria. These organisms can cause symptoms such as nausea, cramps, diarrhea, and associated headaches.	Soil runoff
Barium	2	2	Increase in blood pressure.	Discharge of drilling wastes discharge from metal refineries; erosion of natural deposits.
Cadmium	0.005	0.005	Kidney damage.	Corrosion of galvanized pipes; erosion of natural deposits; discharge from metal refineries; runoff from waste batteries and paints.
Chromium (total)	0.1	0.1	Allergic dermatitis.	Discharge from steel and pulp mills; erosion of natural deposits.
Copper	1.3	TT[8]; Action Level=1.3	Short term exposure: Gastrointestinal distress Long term exposure: Liver or kidney damage People with Wilson's Disease should consult their personal doctor if the amount of copper in their water exceeds the action level.	Corrosion of household plumbing systems; erosion of natural deposits.

© The Watercourse, 2003

118

Drinking Water Standards II

Contaminant (mg/L)[2]	MCLG[1]	MCL or TT[1] (mg/L)[2] MCL	Potential health effects from exposure above the water	Common sources of contaminant in drinking
Cyanide (as free cyanide)	0.2	0.2	Nerve damage or thyroid problems.	Discharge from steel/metal factories; discharge from plastic and fertilizer factories.
Fluoride	4.0	4.0	Bone disease (pain and tenderness of the bones); Children may get mottled teeth.	Water additive which promotes strong teeth; erosion of natural deposits; discharge from fertilizer and aluminum factories.
Lead	Zero Level= 1.015	TT[8]; Action	Infants and children: Delays in physical or mental development; children could show slight deficits in attention span and learning abilities. Adults: Kidney problems; high blood pressure.	Corrosion of household plumbing systems; erosion of natural deposits.
Mercury (inorganic)	0.002	0.002	Kidney damage.	Erosion of natural deposits; discharge from refineries and factories; runoff from landfills and croplands.
Nitrate (measured as Nitrogen)	10	10	Infants below the age of six months who drink water containing nitrate in excess of the MCL could become seriously ill and, if untreated, may die. Symptoms include shortness of breath and blue-baby syndrome.	Runoff from fertilizer use; leaching from septic tanks, sewage; erosion of natural deposits.
Nitrite (measured as Nitrogen)	1	1	Infants below the age of six months who drink water containing nitrate in excess of the MCL could become seriously ill and, if untreated, may die. Symptoms include shortness of breath and blue-baby syndrome.	Runoff from fertilizer use; leaching from septic tanks, sewage; erosion of natural deposits.
Selenium	0.05	0.05	Hair or fingernail loss; numbness in fingers or toes; circulatory problems.	Discharge from petroleum refineries; erosion of natural deposits; discharge from mines.

Drinking Water Standards III

1. Definitions:
   - Maximum Contaminant Level Goal (MCLG)- The level of a contaminant in drinking water below which there is not known or expected risk to health. MCLGs allow for a margin of safety and are non-enforceable public health goals.
   - Maximum Contaminant Level (MCL)- The highest level of a contaminant that is allowed in drinking water. MCLs are set as close to MCLGs as feasible using the best available treatment technology and taking cost into consideration. MCLs are enforceable standards.
   - Treatment Technique (TT)- A required process intended to reduce the level of a contaminant in drinking water.
2. Units are in milligrams per liter (mg/L) unless other wise noted. Milligrams per liter are equivalent to parts per million (ppm).
3. EPA's surface water treatment rules require systems using surface water or ground water under the direct influence of surface water to (1) disinfect their water, and (2) filter their water or meet criteria for avoiding filtration so that turbidity (cloudiness of water) will at no time go above 5 nephelolometric turbidity units (NTU). Systems that filter must ensure that the turbidity goes no higher than 1 NTU (0.5 NTU for conventional or direct filtration) in at least 95% of the daily samples in any month. As of January 1, 2002, turbidity may never exceed 1 NTU, and must not exceed 0.3 NTU in 95% of daily samples in any month.
4. No more than 5.0% samples total coliform-positive in a month. (For water systems that collect fewer than 40 routine samples per month, no more than one sample can be total coliform-positive per month.) Every sample that has total coliform must be analyzed for either fecal coliforms or *E.coli* if two consecutive TC-positive samples, and one is also positive for *E.coli* fecal coliforms, system has an acute MCL violation.
5. Fecal coliform and *E.coli* bacteria whose presence indicates that the water may be contaminated with human or animal wastes. Disease-causing microbes (pathogens) in these wastes can cause diarrhea, cramps, nausea, head aches, or other symptoms. These pathogens may pose a special health risk for infants, young children, and people with severely compromised immune systems.
6. Lead and copper are regulated by a Treatment Technique that requires systems to control the corrosiveness of their water. If more than 10% of tap water samples exceed the action level, water systems must take additional steps. For copper, the action level is 1.3 mg/L, and for lead is 0.015 mg/L.

Wash It Away

Summary:
Students explore how diseases can be transmitted easily within a population by using glitter to represent common pathogens; and then investigate hand-washing as a method of disease prevention.

Objectives:
Students will:
- identify common sources of disease transmission.
- recognize that infectious microbes can be found in common places.
- recognize how easily infection and disease can be spread by contact.
- describe simple methods for preventing transmission of disease and infection.

Materials:
- *Water*
- *Clean spray bottle*
- *Soap* (for hand washing)
- *Glitter*
- *Optional: Treats (candy, fruit, etc. for all students)*

Background:
The human body is home to all kinds of bacteria. Most bacteria are beneficial. In fact, if it weren't for bacteria, people could not exist. Microorganisms digest food and produce substances such as vitamin K that help blood to clot. They form the body's first line of defense by out-competing pathogens (harmful viruses or bacteria) on the skin or in the

Grade Level:
6-8

Subject Areas:
Health Science, General Science, Microbiology

Duration:
Preparation:
20 minutes

Activity:
50 minutes

Setting:
Classroom

Skills:
Model, Cause and Effect, Interpret, Research

Vocabulary:
pathogen, microbe, waterborne

mouth. However, when people get infected with an overwhelming number of pathogens that settle in their nasal passages, throat, lungs, stomach and other parts of the body, they can easily infect others by coughing, sneezing, or through liquids passing out of their bodies. The spread of germs from one person to another is the basis of infection by communicable diseases.

Fortunately, our bodies have many ways to combat infection. Nasal passages and the trachea are lined with tiny hairs that trap microorganisms. Once the body has been invaded by pathogens, it can mount various attacks. Certain white blood cells can seek out pathogens and consume them. Other white blood cells produce antibodies that fit the structure of pathogens much like puzzle pieces. When an antibody finds a pathogen, it attaches to it and destroys it. This requires some time, which is why it can take a week to recover from a cold, influenza, or stomach infection.

Parents, teachers, and the medical community stress the importance of preventive measures to avoid the spread of pathogens and help us stay clear of infection from some diseases. Preventive measures include the following:
- Covering your mouth with your elbow (rather than your hand) when coughing or sneezing
- Washing hands after using the bathroom and before each meal (*See sidebar*)
- Avoiding shared eating utensils and beverages

Simple practices can prevent the spread of pathogens to others while preventing us from becoming infected. With many of these pathogens being waterborne or transmitted through human fluids (e.g., coughing, sneezing, etc.), it is important to understand how easily they are transmitted to others and how they can be prevented.

The United States Center for Disease Control–National Center for Infection Diseases offers the following guidelines for disease prevention:
"The most important thing you can do to keep from getting sick is to wash your hands. In addition to colds, some pretty serious diseases—like hepatitis A, meningitis, and infectious diarrhea—can be easily prevented if people make a habit of washing their hands."

Procedure

Warm Up
Demonstrate how pathogens are passed along to others through touch. Fake a sneeze, covering your mouth with your hand. Wet that hand with a spray bottle of water. Shake hands with a student. Instruct that student to shake hands with the next student without drying their hands. Continue this until all the students have shaken hands.

How many students shook a wet hand? What if pathogens were in the "sneezed" fluids? Discuss how easily microorganisms could get from a person's hand to their

mouth. Ask students to brainstorm better ways to deal with a sneeze than catching it in their hand (e.g., handkerchief, sneeze into their elbow, etc.).

The Activity:
Part I
Note: Use either A, B, or both to demonstrate the process by which pathogens can be transmitted in a population.

A: The Handshake
1. Explain the object of the activity: **The students are to shake hands with as many different students as they can in one minute.**
2. **In secret, one student is selected as the disease carrier and is given a damp washcloth and a bag full of glitter.** Before shaking hands, this student is to stealthily moisten their hand and stick their hand in glitter so that the glitter is transmitted to the other students during a handshake.
3. **Conduct the activity for one minute.** After one minute, ask students to report how many hands they had shaken.
4. **Explain that there was one infected person who was spreading a disease because their hands were not clean. Now ask the students to check their hands to see if they have glitter on them.** The glitter represents microbes that could cause infection. How many students are now carrying the disease (glitter)?

"An Ounce of Prevention: Keeps the Germs Away" (CDC, 2000)

When should you wash your hands?
- Before, during, and after you prepare food
- Before you eat, and after you use the bathroom
- After handling animals or animal waste
- When your hands are dirty
- More frequently when someone in your home is sick

What is the correct way to wash your hands?
1. First wet your hands and apply liquid or clean bar soap. Place the bar soap on a rack and allow it to drain.
2. Next rub your hands vigorously together and scrub all surfaces.
3. Continue vigorously rubbing your hands together for at least 15 seconds. It is the soap combined with the scrubbing action that helps dislodge and remove pathogens.
4. Rinse well and dry your hands.

B: The Handouts
1. "Contaminate" several classroom objects by coating them with glitter, representing disease-carrying microbes.
- Wet the outside classroom door handle and coat with

glitter. Close the door so that students must open it to enter the classroom.

- Coat your hand with glitter and shake hands with students as they arrive.
- **Continue to "contaminate" your hand and distribute pencils, handouts, or other items that become contaminated with glitter.**

Part II

1. After you have "contaminated" the class with glitter, offer the students the opportunity to have a treat or desirable item (e.g., fruit, candy, etc.).
2. **Explain that the glitter represents microbial infections that can potentially cause serious illness.** The glitter started with one person, and was spread by contact to most of the students in one class period.

Puerto Rican boys in water which could contain the agent of bilharzia (schistosomiasis). Courtesy: Centers for Disease Control

3. **Instruct five of the "infected" students to wash their hands with soap and water, using the recommended hand washing guidelines from the Centers for Disease Control.**
4. **Instruct five more of the infected students to quickly rinse their hands with cold water.** Did the hand-washing affect the amount of glitter on their hands?
5. **Ask the students how the transmission of the microbes could have been prevented.**
6. **Ask how many of the students still want to eat their treat?**

Wrap Up:

1. Discuss what diseases could be represented by the glitter and their impacts on human health.
2. Brainstorm methods for preventing disease, focusing on simple ways to reduce disease transmission—hand washing, safe swimming, covering mouth when coughing or sneezing, etc.
3. Create a classroom chart of simple prevention methods. Have students complete the chart, recording the preventive methods that they use each day.
4. Have students choose a water-borne disease (local, national, or international) and research it. Instruct them to determine where it originated, where it can be found today, how it is transmitted, what the symptoms are, what is being done to prevent this disease.

Assessment:

Have students:

- demonstrate how microbial infection spreads by contact (*Warm Up, Part I and Part II*).
- identify simple practices for disease prevention (*Warm Up, Part II*)
- investigate local water health issues (**Activity, Extensions**).

Extensions:

1. Peel 3 potatoes and prepare them according to the following:
a. Have all students touch the first one.
b. After washing their hands using CDC guidelines, have all students touch the second one.
c. Teacher washes the third one with no students touching it.
2. After completing the above steps, seal the 3 potatoes in separate plastic bags and place them in the classroom in a warm location.
3. Periodically over several weeks, have students compare the microbial growth between the potatoes. Have them brainstorm the best methods to measure the amount of microbial growth (without opening the sealed bags), and in small

groups, have them create graphs and charts for comparison.

4. What accounts for the differences in microbial growth on the 3 potatoes? How could some of the growth be prevented? How does this investigation relate to methods that students can practice to prevent the spread of diseases? After the experiment is completed, pour bleach into the bags with the potatoes to kill the bacteria, seal them again, and dispose of them in the garbage or landfill.

Have a guest speaker address the class about a local water-related human health issue. Invite your local health department, an epidemiologist, or doctor to present the impacts and prevention of the local water-related human health issue (e.g., bacterial contamination, beach closures, sewer overflows, drinking pool water, eating contaminated fish, swimming in unsafe waters, etc.).

In small groups, have students write newspaper articles about a local water health issue, presenting a balanced look at the problem and highlighting ways to prevent the disease. Research methods may involve Internet research, outreach to local experts and the medical community, and interviews, much as a journalist would do.

 Testing Kit Extensions:
Test local waters for bacteria using the Healthy Water, Healthy People Hach Paddle Testers. A presence or absence of bacteria will be determined, with a relative quantification of their abundance. Compare water from ponds, rivers, wells, and more. Be certain to follow the procedures for proper handling, treatment, and disposal of the used containers (fill with bleach and dispose of in garbage/landfill).

Have students compare the bacteria from washed and unwashed hands. Using the Healthy Water, Healthy People Hach Paddle Testers, have one group touch the agar on both sides of the paddle with unwashed hands. Have the other group touch both sides of the paddle after washing their hands according to the CDC guidelines. Incubate the paddle testers according to the procedures, or simply keep them on a heater or other warm place for 48 hours. Compare the results of the two groups. Dispose of the paddle testers properly by filling them with bleach and disposing of them in the garbage or landfill.

Resources:

American Water Works Association. 1999. *Manual of Water Supply Practices: Waterborne Pathogens.* Denver, CO: American Water Works Association.

Centers for Disease Control and Prevention (CDC). April 5, 2000. *An Ounce of Prevention: Keeps the Germs Away.* Retrieved February 26, 2003 from the Web site: http://www.cdc.gov/ncidod/op/handwashing.htm

Dixon, B. 1994. *Power Unseen: How Microbes Rule the World.* New York, NY: W. H. Freeman and Company.

Gleick, P.H. 2000. *The World's Water 2000-2001: The Biennial Report on Freshwater Resources.* Washington, D.C.: Island Press.

The Watercourse. 1995. *Project WET Curriculum and Activity Guide.* Bozeman, MT: The Watercourse.

Notes:

Life and Death Situation

Summary
Students learn about the diversity and global locations of waterborne diseases and the role of epidemiology in disease control by searching for others who have been "infected" with the same waterborne illness as they have. Then, they create newspaper articles that give an overview of their disease.

Objectives
Students will:
- compare symptoms of waterborne diseases.
- analyze and plot on a map the transmission of these diseases around the world.
- write a newspaper article describing the location, impacts, and background of waterborne diseases.
- identify the role of water in transmitting diseases.

Materials
- **Symptoms of Diseases** *Student Copy Page* (Make at least 2 copies of the **Symptom Cards** pages; cut and place them into separate envelopes, one disease per envelope. Make sure the order of **Symptom Cards** is different for envelopes containing the same disease and that the name of the disease is not included with the symptom cards. If more than 24 students in class; make extra sets.)
- *Copies of **Disease Descriptions***

Grade Level:
6–12

Subject Areas:
Health Science, Microbiology, Environmental Science, Geography, Biology, Language Arts

Duration:
Preparation:
30 minutes

Activity:
Part I:
50 minutes

Part II: 50 minutes

Setting:
Classroom

Skills:
Gather, Organize, Analyze, Infer, Draw Conclusions

Vocabulary:
waterborne disease, epidemiologist, pathogen, bacterium, protozoan, virus, water treatment

(1 per group)
- *Copies of **Waterborne Disease Worksheet** Student Copy Page* (1 per group)
- *Pencils and note pads (optional)*
- *World Map*

Background
Waterborne diseases are acquired through the ingestion of contaminated water. About 80 percent of all diseases are water-related. In many of these illnesses, water infiltrated with sewage spreads the disease. Also an infected person or animal may pass pathogenic bacteria, viruses, or protozoa through their waste into the water supply.

Because microorganisms that cause illness often cannot be seen, smelled, or tasted, contaminated water often appears fresh and clear. This particularly concerns those responsible for municipal drinking water supplies because contamination often goes undetected until a noticeable number of people become ill.

Most ailments caused by ingesting sewage-contaminated water are intestinal and lead to symptoms like gas, nausea, cramping, and diarrhea. Some pathogens (harmful microorganisms that cause illness) attach to intestinal linings and produce toxic materials that the body then tries to purge. Others invade intestinal epithelial cells and cause inflammation, but do not produce toxins. Fluids containing disease-fighting white blood cells are secreted in the intestine to aid in attacking or flushing the harmful organisms from the body. Unfortunately, the resulting loss of fluids (diarrhea) also causes dehydration, the major concern for patients with these types of diseases.

Dehydration can be life threatening, especially if the patient is very young, elderly, malnourished, or immunocompromised (e.g., those with AIDS or on high doses of antibiotics). Children who get diarrhea should be closely monitored because their immune system is not as developed as an adult's is. The same number of pathogens that can quickly overwhelm a child's system might not harm a healthy adult. As many as one-third of pediatric deaths in developing countries are attributed to diarrhea and the resulting dehydration. Africa, Asia, and Latin America experience an estimated 3-5 billion cases of diarrhea, with 5-10 million attributable deaths, each year. Among the leading causes of bacterial diarrhea are the bacteria, *Vibrio cholerae* (cholera), *Salmonella sp.* (salmonella), *and Shigella* (shigellosis).

Bacteria live everywhere, including our drinking water. However, water supplies are monitored to detect pathogens in infectious concentrations. Water treatment facilities in the U.S. routinely use indicator bacteria levels to watch for these pathogens. By checking levels of indicator bacteria called coliforms (such as *Escherichia coli* –a common organism in our intestines), epidemiologists and other scientists can tell when fecal matter or other pathogens have contaminated the water. Though the presence of coliforms in our water does not mean other pathogens are present, it does indicate the possibility that they may be. Facilities can then accelerate water treatment procedures, while monitoring more intensively to determine if fecal pathogens are present. If so, the source of contamination must be found and protective measures taken to stop further contamination. An example of a protective measure might be to require all cooking or drinking water to be boiled.

Until recently, Americans have regularly suffered through epidemics of waterborne illness such as cholera and typhoid fever. Improvements in wastewater treatment and disposal practices and the development, protection, and treatment of water supplies have significantly reduced the incidence of these diseases. The treatment and disinfecting of municipal water supplies have made infection by microorganisms rare in developed countries; however, treatment of drinking water and wastewater is minimal or nonexistent in many developing countries. In some cases, sewage and other wastes are dumped directly into rivers that provide people downstream with water for drinking and washing.

When outbreaks of a particular disease occur, epidemiologists research symptoms, incidence, and distribution of the cases to try to determine the cause of the disease

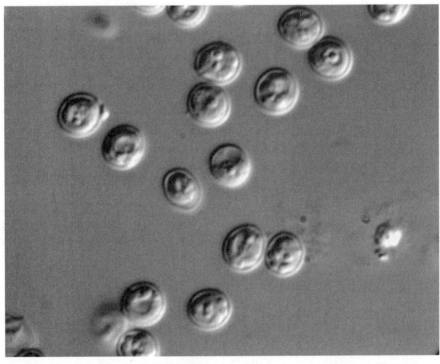

Cryptosporidium magnified 1000x. Cryptosporidium has caused severe illness in some U.S. cities. American Water Works Association

and its means of transmission. This information can help doctors, researchers, and engineers control the disease and aid in preventing it in the future. With waterborne diseases, it is critical to determine how the water supply was contaminated. The case histories of affected patients and any associations among these patients help epidemiologists solve the mysteries of disease.

Procedure

Warm Up

Ask students if they can identify the world's number one cause of death. Explain that millions of people die each year from dehydration caused by diarrhea, which can be caused by a variety of bacteria, viruses, and protozoa. Ask students to brainstorm ways these disease-causing microbes are spread. What is the primary way these organisms are spread? (They are spread primarily by water.)

(Option: Show students two glasses of water: one murky with sediment, the other clear—but possibly containing pathogens. Ask them, "Which glass of water would you prefer to drink?" No matter which one they choose, make the point that disease-causing organisms can be found in clear, "clean-looking" water as well as the "dirty" water.)

The Activity

Part I

1. Tell students that, like epidemiologists, they are going to compare symptoms and modes of transmission of diseases that they have "acquired."

When disease-causing bacteria are found near shell fish beds, harvesting is curtailed. Courtesy: National Atmospheric and Oceanographic Administration

Inform them that they will be searching for others in the class who exhibit similar symptoms of illness.

2. Hand out the symptom envelopes with the *Disease Descriptions* cut individually, one to each student.

3. Direct students to select one clue randomly from the envelope. Have them circulate throughout the room and ask others about their symptoms. Their goal is to locate other students who have similar symptoms. Students may take notes if needed, or this can be a memory exercise. Students can have fun acting out certain symptoms.

4. After a minute or two, tell students to remove a second symptom card from their envelopes. They should continue to

search for others with the same illness.

5. Continue this process until all clues have been removed from the envelopes and everyone has found at least one other person sharing the same waterborne disease.

6. Have students form small groups with others who share the same disease.

Part II

1. Hand out a copy of the *Disease Descriptions* to each group. Have students read their list of symptoms and identify their illness from the *Disease Descriptions*.

2. Have students research their disease and complete their *Waterborne Disease Worksheet* as a group.

3. Those students in the control group who had no illness can choose a disease to research.

4. After completing the questions on the *Waterborne Disease Worksheet,* have groups write a short newspaper article highlighting their disease, its global location, its symptoms, major outbreaks, the number of people it has killed, and other important facts. To provide a guideline, have them answer the five W's (what, where, who, why, and when) and how.

5. Have students share their newspaper articles with the class, or compile them into a newspaper.

Wrap Up

Discuss the role of water in the transmission of disease. Emphasize that most waterborne diseases are a result of inadequate water treatment and poor sanitation. However, contamination does occasionally occur despite sound water treatment practices. This was the case in Milwaukee when an outbreak of *Cryptosporidia* sp. was linked to a contamination of the city's drinking water. This outbreak was particularly disturbing because the city's modern filtration systems failed to remove the extremely small *Cryptosporidia* sp. organism.

Ask students if they know of any cases of hikers or travelers becoming ill after drinking water they thought was safe. Ask students to

brainstorm possible causes of the illness. Discuss how epidemiologists study the cause and the transmission of waterborne diseases.

Assessment

Have students:

- describe the symptoms of several waterborne diseases (*Part I,* Steps 3-6).
- analyze and plot on a map where these diseases occur around the world (*Part II,* Step 2).
- write a newspaper article describing the location, impacts, and background of waterborne diseases (*Part II,* Step 3).
- identify the role of water in transmitting diseases (*Wrap Up*).

Extensions

Have students build on their newspaper articles to develop a multi-media presentation on their diseases to present to the class.

Contact a local health department official, doctor, microbiologist, or water quality specialist and invite them to present local research on waterborne diseases.

Visit your local water treatment plant to learn how they test for waterborne diseases and what steps they take to ensure that drinking water is safe.

Testing Kit Extensions

Using the Healthy Water, Healthy People Paddle Testers, have students test samples of local waters (pond, stream, lake, river, puddles) for bacteria.

Resources

American Water Works Association (AWWA). 1999. *Waterborne Pathogens – Manual of Water Supply Practices.* Denver, CO: American Water Works Association.

Craun, G. 1990. *Waterborne Disease Outbreaks: Selected Reprints of Articles on Epidemiology, Surveillance, Investigation, and Laboratory Analysis.* Cincinnati, OH: Health Effects Research Laboratory, Office of Research and Development, U.S. Environmental Protection Agency.

Gleick, P.H. 2000. *The World's Water 2000-2001: The Biennial Report on Freshwater Resources.* Washington, DC: Island Press.

National Geographic. 2002. *War on Disease.* (201) 2: pp. 5-31.

For information about occurrences of specific diseases within the United States and around the world, students can contact the Center for Disease Control, 1600 Clifton Road NE, Atlanta, GA 30333. (404) 639-3311. www.cdc.gov. The World Health Organization Web site contains information about global health issues, including waterborne diseases www.who.org.

Common Waterborne Diseases and Symptom Cards

Typhoid fever–caused by *Salmonella typhi* **bacteria.**
Case background:

Symptoms occurred ten days after attending a family camp
Discovered that camp's sewage system was faulty and chlorinator was not functioning
Family reunion had used same camp the previous week; two people who attended had recently recovered from typhoid fever
Began feeling lethargic with general aches and pains
Malaise (general weakness and discomfort) and anorexia–loss of appetite
Developed high fever, became delirious
Tender abdomen with rose-colored spots on skin

Legionnaire's disease–caused by *Legionella pneumophilia* **bacteria.**
Case background:

Chain smoker living in warm climate
Lives in a home that is constantly air-conditioned during summer months
Sudden onset of fever that progressed to a high fever with shaking chills
Developed a cough and excessively rapid breathing
Pain in chest; lungs have rattling sound when breathing
General, diffuse muscular pain and tenderness
Intense headache and mental confusion

Cholera–caused by *Vibrio cholerae* **bacteria.**
Case background:

Recently returned from Bangladesh
Symptom occurred two days after eating fruit thoroughly washed at outdoor pump
Family members have begun coming down with the same symptoms
Severe dehydration
Painless diarrhea, vomiting
Severe muscular cramps in arms, legs, hands and feet
Eyes and cheeks appear sunken; hands have wrinkled appearance

Amebiasis–caused by *Entamoeba histolytica*, **a protozoan.**
Case background:

Returned from Thailand two weeks ago
Drank unbottled water
Feverish
General abdominal discomfort and tenderness, especially on lower right side
Dysentery
Tires easily, mental dullness
Moderate weight loss

Common Waterborne Diseases and Symptom Cards

Enterotoxigenic *E. coli gastroenteritis*–**caused by** *E. coli* **bacteria.**
Case background:

Just returned from visiting friends in Mexico
Symptoms began 12 hours after drinking several swallows of water from a bucket pulled from a well
Experiencing dehydration caused by diarrhea
General, diffuse muscular pain and tenderness
Low-grade fever
Nausea, vomiting
Abdominal cramps

Giardiasis–caused by the *Giardia lamblia* **protozoan.**
Case background:

Symptoms occurred two weeks after backpacking trip
Filled water bottle with clear, fresh-tasting water from a stream below a beaver dam
Abdominal cramps
Intermittent dysentery (which is greasy and odorous)
Excessive intestinal gas
Malaise–general weakness and discomfort
Weight loss

Salmonellosis–caused by a species of *Salmonella* **bacteria.**
Case background:

Lives on a ranch that raises cattle and chickens
Symptoms occurred 10 hours after drinking from pump outside of barn (ground water may have been contaminated by surface water in the pasture after heavy rain)
Malaise–general weakness and discomfort
Fever
Dysentery
Abdominal cramps
Nausea and vomiting

Shigellosis–caused by a species of *Shigella* **bacteria.**
Case background:

Four years old
Symptoms began 15 hours after bobbing for apples in pre-school class
Severe abdominal cramps
Frequent, painful dysentery
Blood and mucus in stool
High fever, chills
Dehydration

Common Waterborne Diseases and Symptom Cards

Hepatitis A.–caused by *Hepatitis A* **virus.**
Case background:

Visited favorite beach and swam with friends
Malaise–general weakness and discomfort
Anorexia–loss of appetite
Fever
Nausea, mild diarrhea
Jaundice–yellowing of skin and whites of eyes
Sick for a week

Control Card. (Individual has no waterborne illness.)
Case background:

Lives in an apartment in the city
Chain smoker living in a warm climate
Drinks tap water
Pain in chest; lungs have rattling sound when breathing
Visited favorite beach and swam with friends
Recently visited an alligator farm
Eats lots of fresh seafood

Cryptosporidiosis–caused by *Cryptosporidium.*
Case background:

Four years old
Attends a daycare center five days a week
Diarrhea
Nausea and vomiting
Fever
Sucks thumb
Recently swam in a local pond

Control Card. (Individual has no waterborne illness.)
Case background:

Lives on a ranch that raises cattle and chicken
Just returned from visiting friends in Mexico
Lives in a home that is constantly air-conditioned during summer months
Is tired in the late afternoon
Often conducts pack trips in the mountains
Works 14 hours a day, usually seven days a week
Drinks eight glasses of water per day

Disease Descriptions

Typhoid fever–caused by *Salmonella typhi* **bacteria.**
Now uncommon in the U.S., this is usually acquired during foreign travel. During the first half of this century it was the most commonly reported cause of waterborne disease in the U.S. It can be acquired by contact with contaminated water, swimming, etc. In 1907, Mary Mallon, nicknamed "Typhoid Mary," was identified as a carrier of the disease. She transmitted the disease while working as a cook in restaurants and private homes in New York City. She escaped authorities for eight years, but was finally apprehended in 1915. She infected some 50 people, three of whom died. In 1973, a major outbreak of typhoid fever affected 225 people in a migrant labor camp in Dade County, Florida. The well that supplied water to the camp was contaminated by surface water. Symptoms include tender abdomen, pink spots on skin, high fever. Those infected may also feel delirious and weak.

Legionnaire's disease–caused by *Legionella pneumophilia* **bacteria.**
Found naturally in water environments; bacteria often colonize artificial water systems such as air conditioners and hot water heaters, and can be inhaled with aerosols produced by such systems. Smoking and lung disease increase susceptibility to disease. A person infected may experience rapid breathing, heavy cough, headache, and shaking chills. They may also feel as if their lungs rattle when they breathe.

Cholera–caused by *Vibrio cholerae* **bacteria.**
Extremely contagious; if untreated, dehydration can lead to death. Cholera originated in Europe and was spread to the United States by transatlantic liners through New Orleans. It reached California through the Forty-niners in their quest for gold. Recent outbreaks of cholera have occurred throughout the United States. Along the Gulf Coast, water and seafood were identified as contributing to the outbreaks. In Louisiana, undercooked crab was the culprit. In 1981, people in Texas were infected after eating cooked rice that had been washed with contaminated water. Cholera causes severe muscle cramps in the extremities of the body, dehydration, diarrhea and the infected person's eyes and cheeks may appear sunken in.

Amebiasis–caused by *Entamoeba histolytica,* **a protozoan.**
Usually occurs in tropical areas where crowding and poor sanitation exist. Waterborne outbreaks are now rare in the United States. The *Entamoeba histolytica* protozoan will cause abdominal tenderness and discomfort, dysentery, and fever.

Enterotoxigenic *E. coli gastroenteritis*–**caused by** *E. coli* **bacteria.**
Leading cause of infant mortality worldwide. Visitors to Latin American countries who partake of the food and water occasionally come down with "traveler's diarrhea," also known as "turista" or "Montezuma's Revenge." A large outbreak of this disease occurred in 1975 in Crater Lake National Park, Oregon. About 2,000 park visitors and 200 park employees became ill after consuming water contaminated by sewage. Abdominal cramping, nausea, and vomiting are all tell-tale symptoms of the infection.

Giardiasis–caused by the *Giardia lamblia* **protozoan.**
Sickness results with only a low dose of the protozoan; it is the most commonly reported cause of waterborne outbreaks. The *giardia* protozoan is killed by boiling water for at least five minutes or is removed by passing water through a filter whose pore size is 0.2 microns or smaller. Infected persons will experience weight loss, intestinal gas, and intermittent dysentery.

Disease Descriptions

Salmonellosis–caused by a species of *Salmonella* bacteria.
Carried by humans and many animals; wastes from both can transmit the organism to water or food. The largest waterborne salmonella outbreak reported in the United States was in Riverside, California, in 1965 and affected over 16,000 people. Salmonellosis causes fever, nausea, vomiting, and dysentery.

Shigellosis–caused by a species of *Shigella* bacteria.
Most infection is seen in children 1-10 years old; a very low dose can cause illness. Waterborne transmission is responsible for a majority of the outbreaks. Children infected may experience blood and mucus in stools, frequent and heavy dysentery, and high fever.

Hepatitis A.–caused by *Hepatitis A* virus.
It is the third most common cause of waterborne disease in U.S. The term "hepatitis" refers to inflammation of the liver. Hepatitis A causes jaundice (a yellowing of the skin and whites of the eyes), nausea and fever, and lasts approximately one week.

Cryptosporidiosis–caused by *Cryptosporidium.*
This was first identified as a cause of diarrhea in people in 1976. It can be transmitted through contact with animals (particularly cattle and sheep), other humans (especially in daycare centers), and contaminated water supplies. Infection will cause diarrhea, fever, nausea, and vomiting.

Waterborne Disease Worksheet

Now that you have identified others with an illness similar to yours, work as a group to research your disease and answer the following questions:

Disease Name: _____

1. How did you contract the disease?

2. Where does your disease occur in the U.S. and the world? Plot the disease on a world map.

3. How is your disease transmitted to others?

4. How can the spread of your disease be prevented?

5. List conditions that might help your disease spread (e.g., inadequate water treatment systems, concentrated population, political upheaval that forces large migrations of people, organisms such as beavers or snails, etc.).

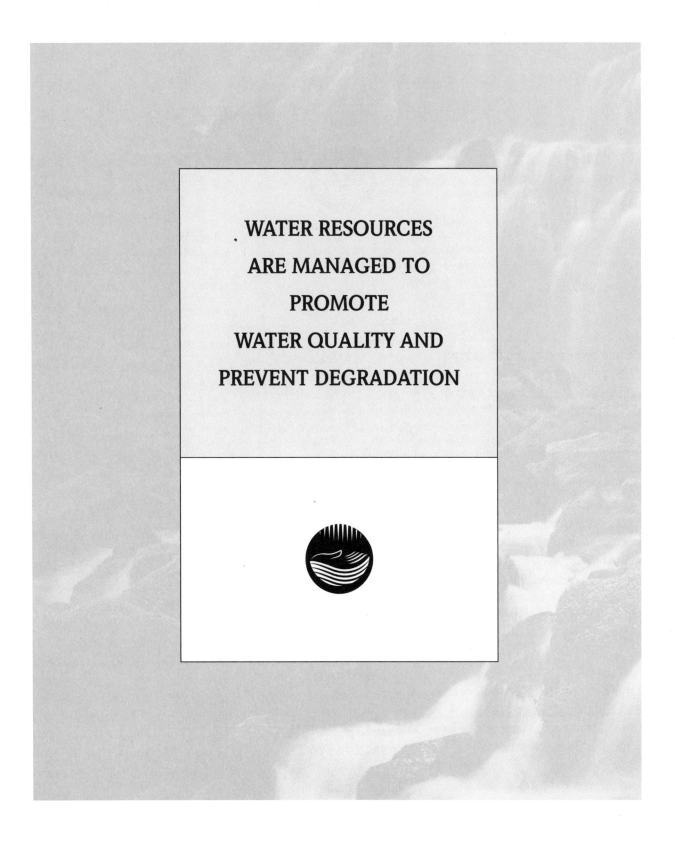

WATER RESOURCES
ARE MANAGED TO
PROMOTE
WATER QUALITY AND
PREVENT DEGRADATION

There Is No Point To This Pollution!

Summary
Students analyze data to solve a mystery, interpret a topographic map, and analyze and compare water quality data to learn about the cumulative impacts of nonpoint source pollution.

Objectives
Students will:
- identify point and nonpoint source (NPS) pollutants.
- demonstrate the cumulative effects of NPS pollution.
- learn to read and interpret a contour map while identifying important map clues about watersheds and water quality.
- graph, analyze, and interpret data sets to draw conclusions about a pollution source.
- compare local household and community NPS pollution to surface water quality standards.
- list ways to reduce or eliminate NPS pollution.

Materials
- *Large clear bowl, 2-liter bottle with top cut off, jar, or baking dish*
- *Water* (enough to fill large bowl or bottle plus 25 ml per student)
- *Clear plastic cups* (1 per student; fewer if used as instructor demonstration)
- *Assorted food coloring* (dilute 1 drop in 250 ml of water for each color)
- *Powdered cocoa or hot chocolate*

Grade Level:
6-12

Subject Areas:
Environmental Science, Geography, Mathematics, Ethics, Social Studies, Language Arts

Duration:
Preparation:
20 minutes

Activity:
two-three 50–minute periods

Setting:
Classroom

Skills:
Graph, Interpret, Debate, Quantify, Compare, Analyze

Vocabulary:
nonpoint source pollution (NPS), point source pollution, runoff, sediments, nutrients, cumulative, turbidity, dissolved oxygen, watershed

mix (14.8 ml or 1 tablespoon)
- *Cooking oil* (5 drops)
- *Eyedropper* (optional)
- *Copies of **Loop Lake Worksheet** Student Copy Page* (8–one per group)
- *Copies of **Loop Lake Map** Student Copy Page* (8–one per group)
- *Copies of **Water Quality Data** Student Copy Page* (8–one per group)
- *Copies of **Water Quality Graphs** Student Copy Page* (8–one per group)
- *Optional:* Local (state) Surface Water Quality Standards, available from: http://www.epa.gov/ost/standards/wqslibrary/index.html

Background
Oil spills, leaking toxic waste, and industrial smokestack emissions make headlines in the news. These kinds of environmental tragedies, where the source of pollution is an obvious known point, are called point source pollution. Though these pollution events are highly visible, the truth is that the "nation's leading source of water quality degradation is nonpoint source (NPS) pollution, where there is no single point of pollution" (United States Environmental Protection Agency). NPS reaches our waterways via runoff from rainfall and snowmelt.

As runoff travels over land, it picks up soil particles as well as natural and human-made pollutants that eventually end up in rivers, lakes, and oceans. While some of these contaminants are benign as they reside in the soil, they can become problematic when carried into waterways via runoff. Common types of NPS pollution include sediments, nutrients, and contaminants such as pesticides and petroleum products.

Sediments can come from construction sites, poorly managed farmlands, logging sites, and eroding stream banks. Although sediments do not seem like a pollutant, they can cause damage to aquatic organisms by clogging the gills of fish, filling in pore spaces between rocks where macroinvertebrates live and fish lay their eggs, and harboring nutrients such as phosphates and nitrates that can cause algae blooms.

Sediments also increase turbidity, allowing less sunlight to penetrate the water, which harms plant and animal life. This increased sediment influences turbidity, which is "an optical property of water based on the amount of light reflected by suspended particles" (USEPA, 1999). Thus, very turbid water appears murky or cloudy. To measure the turbidity, samples are collected and measured using a nephelometer, or turbidimeter, which uses a light beam and a photoelectric cell to electronically measure the amount of light scattered by suspended particles. The units of measurement are called Nephelometric Turbidity Units or NTUs. The higher the NTU (turbidity) value, the more light is scattered and, generally, the more solids it contains.

To decrease the amount of sediment reaching our waterways and prevent unnecessary erosion, farmers use techniques such as contour tilling, loggers leave buffer zones of trees along rivers, lakes and oceans, and builders and homeowners plant vegetation along stream banks. All natural waters are somewhat turbid, even if only at microscopic levels.

Excessive nutrients can be a result of improper disposal of human and animal/pet waste, as well as runoff from farmlands or lawns and gardens. Human and animal/pet wastes contain nitrates and phosphates. When these wastes are improperly managed, they can wash into storm drains and surface water during rainstorms.

A common but sometimes overlooked source of nutrients in surface water comes from fertilizers applied by homeowners to their lawns and gardens. Many homeowners often over-fertilize to keep their lawns and gardens fertile and green. When it rains, excess fertilizers can run into streams and storm drains (which often drain directly to a stream or

Storms can erode soil directly into storm drains, which often flow directly into nearby streams. Courtesy: United States Environmental Protection Agency

river). Runoff from farmlands can carry nitrate-rich fertilizers into surface water, especially if fertilizers have been over-applied or haven't fully infiltrated the soil before a heavy rain comes. Applying fertilizers in the minimum amount necessary, or fertilizing with compost made from recycled animal waste help control nutrient loads.

However common it may be, nonpoint source pollution is not front-page news. Rarely does a newspaper headline read, "Neighbor cited for dumping used motor oil down the drain! Read all about it!" Because we all contribute to the problem, we can all contribute to the solution as well. By taking simple actions like using caution when changing motor oil or filling the lawnmower with gas, planting buffer strips along waterways and drainage areas, and applying fertilizers as instructed, we can minimize the amount of NPS pollution in our waterways.

Procedure
Warm Up
Part I

1. Lead a discussion on pollution. Ask students to brainstorm a list of pollutants while you write them on the board.

2. Explain the difference between point source pollution (pollution source is a known point–e.g. an effluent pipe from a factory), and nonpoint source pollution (pollution source is not defined by a point, also called runoff–e.g., oil and gas from city streets).

Algal bloom on a lake caused by phosphates. American Water Works Association

3. Go through the list and have students identify whether each pollutant is classified as a point source or nonpoint source pollutant.

Part II

Note: Also works well as an instructor demonstration.

1. Provide each student with approximately 50 ml of clear water in a small cup.
2. Each student represents one household and the water in their

cup represents the water that flows across their property. Students should choose one pollutant for their household from the list in *Part I.*
3. Show students the pollution indicators that will represent the pollution flowing across their property. The diluted food coloring will represent fertilizers; powdered cocoa will represent sediment; cooking oil will represent used motor oil, etc.
4. Have each student place a small amount (e.g., 1 drop of diluted food coloring; pinch of cocoa; drop of cooking oil; etc.) of pollution into his or her cup according to the pollution indicators provided.
5. Fill the large clear bowl (or clear 2-liter bottle) half-full with clear water and place it in a central location. The bowl represents a lake that is surrounded by the students' homes. One at a time, have students pour their water into the bowl, announcing which pollutants are

present in their cup of water.
6. Notice that though each individual household contributes only slightly polluted water, once the whole neighborhood has added their water, the lake becomes very polluted. Students should be able to see through each of their samples, but the watershed water should be very dark. This illustrates the cumulative effect of NPS pollution and how we all contribute to it.

The Activity

Part I

1. **Divide students into eight groups and distribute the Loop Lake Map and the *Loop Lake Worksheet* to each group.**
2. **Instruct groups to complete Part I: Map Reading of the *Loop Lake Worksheet.* Have them share their answers after completion.** *(Option: Skip Part I if your students are experienced in reading and interpreting a contour map.)*
3. **Encourage students to find the various locations named as you present the following scenario:**
Loop Lake Village, the community surrounding Loop Lake, is small but prosperous. Residents of Loop Lake Village include Jack Pine, the local nursery grower who has just expanded his operation and Mr. and Mrs. Holstein who own and operate a dairy farm with cows and sheep. Around the lake are several small neighborhoods. There is a well-maintained trailer park that has been a stable part of the

community for many years. There is an older housing development with meticulously maintained lawns stretching down to the lake. Idle farmland was recently sold and a new subdivision is under construction in its place.

There is a bustling town center with a strip mall and large parking lot. The town's park has a public swimming pool and a popular pet play area. On the other side of the hill, across Watershed Divide Road, there is a new factory that provides several jobs for local community members.

This summer, for the first time ever, a large algal bloom (when algae grows at a rapid rate) developed in the lake, and several dead fish washed up on the shore. Some speculate that wastes are being dumped from the factory, which are then draining into the lake. The locals feel it couldn't be just coincidence that there is a new factory and now, for the first time, the lake has an algal bloom and dead fish. The town residents agree that in the last ten years the clarity of the lake has decreased, while the temperature and algal growth have increased.

Part II

4. Have students complete Part II of the *Loop Lake Worksheet* and share their answers.
(They should determine that the factory resides in a separate watershed from Loop Lake, and therefore their wastes can not enter Loop

Lake via runoff.)

5. Ask students to listen carefully as you continue reading the scenario.

The factory owners assure the community that they are careful with their wastes and they meet the environmental regulations for hazardous waste disposal. They feel certain they are not harming Loop Lake or the surrounding environment. Understanding that factories can easily be used as scapegoats, factory owners sponsor a surface water quality sampling project to determine who is really polluting Loop Lake. The owners also subsidize purchasing of the monitoring equipment for the sampling.

On the volunteer water sampling day, community members arrive eager to participate. Three parameters will be tested: fertilizers/

nutrients, dissolved oxygen, and turbidity, a measure of water clarity. Samples are collected from streams that flow through various properties surrounding the lake, all of which eventually flow into the lake.

6. Explain to student groups that they will now represent volunteer water quality monitors, each stationed at a particular sampling site. Their challenge is to investigate and determine the cause of the algae bloom and fish kill in Loop Lake by "collecting" and analyzing the water quality data (found on their *Water Quality Data* card). Distribute a copy of their *Water Quality Data* card and inform groups that the data reflects samples taken from the stream running through their site. Discuss Nephelometric Turbidity Units (NTUs) a

Runoff from streets is another source of nutrients entering a stream or lake.
Courtesy: US Environmental Protection Agency

common unit of measure for the turbidity of water. Remind them that the lower the number of NTUs, the clearer the water and the higher the number, the cloudier the water.

7. Explain to students that one way to analyze data is to graph it and compare it with graphs of data for the same parameters. In this activity they will compare their new data with that of Loop Lake today, Loop Lake of ten years ago, state water quality standards, and the data from other groups.

8. Distribute the *Water Quality Graph* to each group and explain that the data for Loop Lake and the state standards are already graphed. Instruct students to graph the data from their sampling site. Tell them to compare their stream data to the other graphs. *(Option: Substitute your state's actual surface water quality standards, if available, for those in the activity. These state-specific standards, if available, can be downloaded from:* http://www.epa.gov/ost/standards/wqslibrary/index.html.)

9. Make an overhead of the *Water Quality Criteria Graphs* or draw the graphs on the board. Have each group plot their data on one main graph to allow all groups to compare their data.

Part III

10. Have students complete Part III of their *Loop Lake Worksheets.*

11. After students share their predictions, ask them what further information may be needed to make a prediction about the cause of the algae bloom and fish kill. After hearing their ideas, share this article printed in the local paper on the day following the volunteer water quality sampling:

After sampling Loop Lake, the Water Quality District has determined that the fish died due to a lack of dissolved oxygen, which they need to live. Dissolved oxygen is forced out of water when algal blooms occur and also when water warms. While algae are typically a source of dissolved oxygen, an algal bloom causes a rapid growth in algae which results in increased die-offs, the end result being a decrease in dissolved oxygen as the bacteria that decompose the dead plants rapidly consume the dissolved oxygen. Nutrients and fertilizers reaching the lake via runoff often cause algal blooms. The blooms then increase turbidity, preventing sunlight from reaching the deeper aquatic plants and killing them. As these plants die, the bacteria that decompose them will consume even more dissolved oxygen as they multiply. Higher turbidity also increases water temperature because the suspended particles absorb heat from the sun, further depleting the dissolved oxygen. We need your help to find out where these fertilizers and nutrients are coming from.

12. Have students analyze the data again, focusing on identifying the cause of the algal bloom and fish kill in Loop Lake based on this recent report. Ask them to complete Part IV of the *Loop Lake Worksheet.*

Wrap Up

Have groups share their answers to the questions in Part IV of the *Loop Lake Worksheet.* The students should conclude that no single source of fertilizers/nutrients exceeds the state standards, but the combined impact of all the sources accumulating in Loop Lake is causing the algal bloom and fish kill.

Lead a discussion using the following questions: Why does the older housing development with big lawns contribute so much fertilizer and nutrients to the lake? (Over-fertilization of lakeside lawns allows fertilizers to run directly into the lake.) Why is the park a heavy contributor of fertilizers/nutrients? (Fertilizers and pet waste contains high levels of nutrients.) How does the new subdivision contribute to the turbidity of the lake? (Construction excavation increases erosion that, in turn, increases turbidity. Phosphates, a limiting nutrient in freshwater systems, can be carried with sediments and released into the water, there by increasing the nutrient load even further.)

Review the causes of the algae bloom and subsequent fish kill. (Accumulated fertilizer/nutrient loading from numerous sources around the lake, increased turbidity and sedimentation, decreased sunlight penetration into the water, accelerated death of deeper water plants, decreased dissolved oxygen levels as bacteria decompose dead plants.)

Divide class into small groups and have each write a short paragraph outlining possible ways each contributor on the list could reduce fertilizers and nutrients deposited into the lake. Discuss their solutions. Ask students to brainstorm ways to reduce household nonpoint source pollutants. Examples: use designated hazardous waste disposal facilities for used motor oil, either reduce the amount of chemical fertilizers used or replace them with natural fertilizers, plant vegetative buffer strips around waterways, clean up pet wastes, and others.

Assessment

Have students:

- identify point and nonpoint source (NPS) pollutants (**Warm Up**).
- demonstrate the cumulative effects of NPS pollution (**Warm Up**).
- read and interpret a contour map while identifying important clues about watersheds and water quality (**Part I**).
- graph, analyze, and interpret data sets to form conclusions about pollution sources (**Parts I II, III**).
- compare community pollution to surface water quality standards (**Part III**).
- list ways to reduce or eliminate NPS pollution from your yard or community (**Wrap Up**).

Extensions

Invite your local water quality district specialist or other water quality expert to discuss which water pollutants are most problematic in your community. What are the sources of these pollutants? Are they nonpoint source or point source pollutants? What is being done in your community to minimize water pollution?

Testing Kit Extensions

Organize a water quality monitoring project for a local watershed. Refer to the study design overview in the activity *Water Quality Monitoring; From Design to Data* on page 70 in this guide.

Using the water quality sampling data, have students calculate the percent saturation of dissolved oxygen in the water. Refer to dissolved oxygen percent saturation charts in the *Healthy Water, Healthy People Instructors Manual* for calculations.

Resources

Goo, R. n.d. *Do's and Don'ts Around the Home*. Retrieved on July 27, 2001, from the Environmental Protection Agency Web site: http://www.epa.gov/owow/nps/dosdont.html

Horton, G. 2001. *Dictionary of Water Words*. Reno, NV: Great Basin Research.

The Watercourse. 1995. *Project WET Curriculum & Activity Guide*. Bozeman, MT: The Watercourse.

United States Department of Agriculture, Natural Resources Conservation Service. 1994. *The Phosphorus Index: A Phosphorus Assessment Tool*. Retrieved on July 23, 2001, from the Web site: http://www.nhq.nrcs.usda.gov/BCS/nutri/phosphor.html

United States Environmental Protection Agency (EPA) a. 1999. *Guidance Manual for Compliance with the Interim Enhanced Surface Water Treatment Rule: Turbidity Provisions*. Retrieved on December 3, 2001, from the Web site: http://www.epa.gov/safewater/mdbp/mdbptg.html

United States Environmental Protection Agency. n.d. *What is Nonpoint Source (NPS) Pollution?* Retrieved on July 27, 2001, from the Web site: http://www.epa.gov/owow/nps/qa.html

United States Environmental Protection Agency. n.d. *Managing Nonpoint Source Pollution from Households*. Retrieved on July 27, 2001, from the Web site: http://www.epa.gov/owow/nps/facts/point10.htm

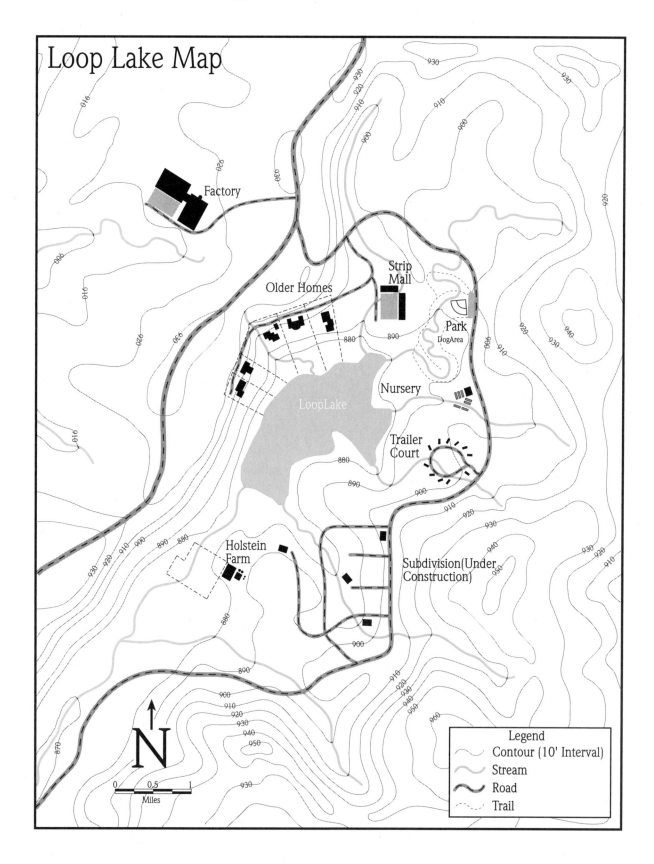

Loop Lake Map

Factory

Strip Mall

Older Homes

Park
DogArea

Nursery

LoopLake

Trailer Court

Holstein Farm

Subdivision(Under Construction)

N

0   0.5   1
Miles

Legend	
	Contour (10' Interval)
	Stream
	Road
	Trail

Loop Lake Worksheet

Part I

Map Reading

1. Locate the contour lines on the map. These lines indicate the elevation of the land at a location on the map. Since there are ten feet between each contour line, comparisons with other elevations can be made. Place an X on the highest and lowest points on the map.

2. Locate the streams on the map. Streams always flow from the highest to the lowest elevations. How many streams are there? _____ Locate the stream that flows through the subdivision and mark its highest and lowest elevations with an X. What is the elevation of the highest point on this stream? How many feet in elevation does this stream drop between its highest and lowest points?

3. Place a Y on the highest points of the streams that flow into Loop Lake.
   a. Place a Z on the highest points *above the streams* that flow into Loop Lake
   b. Following the highest elevations between these points, connect these Zs with lines.
   c. This outline encompasses all of the land area that drains into Loop Lake. What is the name for an area that is drained by streams and flows to a central location? _____

4. One stream serves as the outlet stream for Loop Lake. Locate it and name the business that the stream flows through. How did you know that this was the outlet stream for the lake?

_____

Part II

1. After hearing the story of the fish kill in Loop Lake, write whether you think the factory is to blame. Why or why not?

Part III

1. After analyzing the water quality data, write your predictions of the causes of the algal bloom and fish kill. Be sure to write your reasons or evidence for your prediction.

Part IV

1. How many of the land uses around Loop Lake have fertilizer/nutrient levels that exceed the state standards? _____ For turbidity? _____

2. Write an explanation for how the algal blooms and fish kill occurred in Loop Lake:

3. Is the pollution that caused the fish kill in Loop Lake from point sources or nonpoint sources?

Water Quality Data

Mr. and Mrs. Holstein's Farm Dissolved Oxygen: 8.7 mg/L Turbidity: 12 NTU Fertilizers/Nutrients: 0.3 mg/L Temperature: 15° C (59 F)	**Parking Lot** Dissolved Oxygen: 5.6 mg/L Turbidity: 18 NTU Fertilizers/Nutrients: 0.1 mg/L Temperature: 15° C (59 F)
New Apartments –Construction Dissolved Oxygen: 5.8 mg/L Turbidity: 33 NTU Fertilizers/Nutrients: 0.1 mg/L Temperature: 15° C (59 F)	**Loop Lake Park and Pet Play Area** Dissolved Oxygen: 5.1 mg/L Turbidity: 6 NTU Fertilizers/Nutrients: 0.5 mg/L Temperature: 15° C (59 F)
Factory Dissolved Oxygen: 8.9 mg/L Turbidity: 7 NTU Fertilizers/Nutrients: 0.3 mg/L Temperature: 15° C (59 F)	**Older Housing Development** Dissolved Oxygen: 5.2 mg/L Turbidity: 4 NTU Fertilizers/Nutrients: 0.48 mg/L Temperature: 15° C (59 F)
Jack Pine's Nursery Dissolved Oxygen: 5.7 mg/L Turbidity: 10 NTU Fertilizers/Nutrients: 0.3 mg/L Temperature: 15° C (59 F)	**Trailer Park** Dissolved Oxygen: 5.9 mg/L Turbidity: 9 NTU Fertilizers/Nutrients: 0.3 mg/L Temperature: 15° C (59 F)

Washing Water

Summary

Students model how water travels from its source, through drinking water treatment, contamination, wastewater treatment, and back to the source. Engineering, chemistry, and problem-solving skills are applied as the students design methods for cleaning polluted water.

Objectives

Students will:

- demonstrate the drinking water treatment process using common items.
- create a process for wastewater treatment.
- evaluate the effectiveness of drinking water and wastewater treatment techniques.

Materials

Drinking water treatment materials:

- *1-gallon (3.8 l) jugs of source water* (2)
- *2-liter bottle with cap* (1 per group)
- *Scissors or knife*
- Copies of **Drinking Water Worksheet** *Student Copy Page (1 per group)*
- Copies of **Treatment System Illustrations** *Student Copy Page* (1 per group)
- *Alum* (potassium aluminum sulfate, available in pharmacies, or the spice/baking aisles in grocery stores)

Grade Level:
6–12

Subject Areas:
General Science, Chemistry, Environmental Science, Mathematics, Health Science

Duration:
Preparation:
30 minutes

Activity:
Three to four 50-minute periods

Setting:
Classroom

Skills:
Gather, Measure, Plan, Manipulate, Model

Vocabulary:
aeration, coagulation, sedimentation, filtration, hydrologic cycle, floc, flocculation, source water, disinfection, sludge, potable

- *Plastic spoons* (at least 1 per group)
- *Fine sand* (1 cup per group)
- *Coarse sand* (1 cup per group)
- *Small gravel* (1 cup per group)
- *Large beaker or jar* (1 per group)
- *Coffee filter* (1 per group)
- *Rubber band* (1 per group)
- *Clock*
- *Cardboard box top; lid from a box of paper* (forms a base to support 2-liter bottle)

Wastewater materials

- Copies of **Wastewater Worksheet** *Student Copy Page* (1 per group) (Small amounts or a sample of some of the following):
- *Coffee grounds*
- *Salt*
- *Vegetable oil*
- *Soil*
- *Yeast*
- *Soap*
- *Food scraps*
- *Vinegar*

Possible cleaning materials:

- *Screens to use as filters* (cheese cloth)
- *Coffee filters*
- *Alum*
- *Bowls or cups*
- *Straws or pipettes*
- *Spoons*
- *Baking soda*
- *Charcoal*
- *Talc*
- *Sand and gravel*

Possible testing materials:

- *pH paper (litmus paper)*
- *Brown paper bag*

Background

The process known as the hydrologic cycle describes water's constant movement between oceans and atmosphere, clouds and rivers, ground water and lakes. Several processes of the hydrologic cycle help purify water— evaporation, freezing, and infiltration. During evaporation and freezing, dissolved and suspended particles are left behind as the water changes between states. During infiltration, suspended particles are filtered as water moves through soil, sand, and gravel.

While this system is very effective, it also takes a long time. Because of the human demand for clean drinking water and because humans produce so much wastewater, the hydrologic cycle simply can't keep up. Therefore, we must treat water both before and after we use it. The process of purifying water for human use is not unlike the steps water passes through in the hydrologic cycle. Once source water (water from lakes, ground water, or rivers) reaches a water treatment plant (via pipes), the process begins.

First water is screened to remove large debris. Next, in a step called *pre-chlorination*, chlorine is added to destroy any pathogenic bacteria. In the next step, *flocculation*, alum $(Al_2(SO_4)_3)$ and slaked lime $(Ca(OH)_2)$ are added and react together to form a sticky compound which attracts and clumps suspended particles. These clustered particles, or *flocs*, settle to the bottom of a tank in a process called *settling*, or *sedimentation*. Next the water is filtered through sand and gravel to remove any remaining suspended particles. This is called *sand filtration*. At this point, the water is considered potable (suitable for drinking) but must undergo a final step called *post-chlorination* where chlorine is added to keep the water bacteria-free during delivery from the water treatment plant, through pipes, to homes and businesses. Other optional steps that water treatment facilities can use include aeration to improve the taste of water, fluoridation to reduce tooth decay, and pH adjustment to prevent corrosion of pipes.

After water is used and has disappeared down the drain, it flows through another set of pipes to a wastewater treatment plant. There it again passes through a series of purifying steps designed to make it relatively free of waste before it is returned to a source water system.

Primary wastewater treatment, which involves filtration, settling, and skimming procedures, removes 45-50% of pollutants. Most developed countries apply secondary treatments, mainly biological processes, which remove 85-90 percent of remaining pollutants. Additionally, microorganisms are introduced to the water to consume other organic waste material. Later, in settling

Wastewater treatment facility. American Water Works Association

tanks, solids and microorganisms are separated from the wastewater, which is then disinfected with chemicals like chlorine to kill remaining bacteria. Then it is released into nearby waterways.

Though this process is extensive, some pollutants like nitrates, phosphates, heavy metals, pesticides and cleansers can remain. Some advanced wastewater treatment facilities impose tertiary processes like activated charcoal filtration to remove organic materials, distillation to remove salts, and flocculation to remove suspended matter.

Unfortunately the costs to build and operate such advanced treatment plants are often prohibitive and few communities have them. The expense and difficulty of removing some pollutants means it is best to keep them from reaching water in the first place. Properly disposing of items we use every day like motor oil, gasoline, paint thinner, and pesticides, combined with choosing less caustic cleansers (vinegar, water, baking soda and others) can help keep our water safe.

Procedure

Warm Up

1. Ask students to list the source of their school's drinking water. Is this water treated before it reaches the school? If so, where and how is it treated?

2. Where does their water go after they flush their school's toilets? Where and how is this wastewater treated?

3. Was the school's source water (water that is treated for drinking) previously someone else's wastewater? If a local map is available, have students trace their source water upstream to see if another community is dumping their treated wastewater into it. This is often the case as we "all live downstream" from someone.

4. Explain that we actively treat our water before drinking it to ensure it is potable, and after we use it to ensure that it does not impact the water quality of the waterway that receives it. Inform students that in this activity they will model the processes of water treatment and wastewater treatment.

The Activity

Part I:

Making drinking water Note: Remind students to **NEVER** drink the water during this entire activity!

1. **Preparation: Collect source water (such as that from a river, pond, or lake, or add dirt or mud to tap water) in two 1-gallon (3.8 l) jugs with lids.**

2. **Explain to students that they will attempt to make this source water into drinking water just as a municipal water treatment plant does.** However, since they do not have the proper equipment, remind students to **NEVER** drink this water, even after treatment.

3. **Write the terms Aeration, Pre-Chlorination, Coagulation, Sedimentation, Sand Filtration, and Post-Chlorination on the board or a flipchart. Explain to**

students that these are the primary steps in drinking water treatment. The chlorination steps will be omitted in this activity for safety reasons. Ask students to define these terms as best they can before giving them an overview of the process. The process begins with *Aeration,* which adds oxygen to the water and removes trapped gasses. Next, *Pre-chlorination* destroys pathogenic bacteria. *Coagulation,* which removes suspended solids, is initiated by adding alum to the water. *Sedimentation* allows the floc (coagulated particles) to settle out. *Sand Filtration* removes finer suspended particles. *Post-chlorination* ensures that bacteria will not infect the water as it travels through the distribution system to the consumer. Ozone and ultraviolet radiation are also used in place of chlorine.

4. **Divide students into groups. Distribute a copy of the *Drinking Water Worksheet* and a set of drinking water treatment materials to each group.** Have them cooperatively complete the *Drinking Water Worksheet* in their groups.

5. **Pre-measure a teaspoon (4.9 ml) of alum crystals for each group and carefully distribute it to groups as they request it for their coagulation (Step 3 of the worksheet).** (Advise students not to touch the alum, but if they do, they must immediately wash the affected area with soap and water.)

6. **After the groups have completed their Drinking Water**

Worksheets, have them share their results and discuss the activity. Instruct them to pour their treated water back into the original source water.

Part II: Treating Wastewater

1. Explain to students that after water is purified for drinking, it is distributed to homes and industries where it is used for bathing, washing dishes and clothes, and in manufacturing processes. Pipes then carry wastewater to a wastewater treatment facility, where it is cleaned before it is returned to the source water. Inform students they will be challenged to model this wastewater treatment process using common materials and their own design.

2. Have students "make" wastewater by adding soil, coffee grounds, vegetable oil, soap, salt, vinegar, yeast, and food scraps to the two 1-gallon (3.8 l) jugs of source water.

3. Shake the jugs thoroughly and distribute about .5 liters of the wastewater to each group in the bottom half of their 2-liter bottle.

4. Distribute copies of the *Wastewater Worksheet* and a set of testing materials to each group.

5. Tell students their task is to clean the water. Groups should use their *Wastewater Worksheet* to record their procedures for cleaning the water. Encourage them to use a series of steps, so they

can evaluate each step separately. They may wish to use some of the techniques learned in *Part I,* but are not required to do so.

6. Describe the materials that can be used for the wastewater treatment and instruct students about safety procedures (do not touch the alum or baking soda, and wash their hands or affected areas immediately after touching the wastewater or cleaning agents).

7. After attempting to clean their water, students should evaluate the results. Evaluation criteria may include color, smell, pH, and the presence of oil. Are there any contaminants that they could not remove?

8. After all groups have completed their worksheets, have them share their procedures and results. They should then compare the treatment procedures of all groups and determine which techniques worked well and which did not.

9. Distribute the Wastewater Treatment Illustration to the groups. Review the treatment processes and have students compare their procedures with the actual wastewater treatment process.

Wrap Up

Have students draw or write a description of the process of water treatment from the source water, through drinking water treatment, contamination during use, wastewater treatment, and back to the source water. Challenge them to complete this assessment without

looking at the illustrations provided during the activity.

Ask students to name shortfalls that they noticed in the water treatment system. List them on the board. Balance this list with components of the water treatment system that they see as indispensable. Invite a drinking water and wastewater treatment professional to your classroom to discuss the water treatment process, address the students' questions, and discuss the students' shortfall and indispensable components lists. Have this professional explain why all contaminants are not removed during treatment (money, available resources, time, space, or practicality).

Remind students that the chemicals they pour down their household drain can end up in a wastewater treatment plant, and therefore in drinking water. Have students brainstorm ways to reduce their impact on the drinking water and wastewater treatment systems (e.g., do not pour oil, paints, chemicals down the drains, conserve the amount of water they use and send to these facilities, substitute nontoxic cleaners for chemical cleaning agents, etc.).

Assessment
Have students:
- simulate the process of drinking water treatment using common items (*Activity*-Part I).
- design and implement a waste water treatment process using common items (*Activity*-Part II).

• evaluate the effectiveness of their drinking water and waste water treatment techniques (*Activity*- **Part II**).

Extensions

Contact your local drinking water and wastewater treatment facilities to find out the costs of these processes. Assign costs to the activity above for treatment chemicals, pumping water, cleaning and replacing filters, sludge removal, etc. Conduct the activity again, challenging students to clean their water for as low a cost as possible.

Have students research different ways to reduce their impact on the water treatment systems (e.g., do not pour oil, paints, chemicals down the drains, conserve the amount of water they use and send to these facilities, substitute nontoxic cleaners for chemical cleaning agents, etc). Ask students to present their findings to the class, then plan and conduct a campaign to educate other students and community members about the importance of reducing their impacts on the water treatment system.

Have representatives from a drinking water or wastewater treatment facility, county sanitarian, or water treatment engineer visit the class during the activity to assess the methods for cleaning water used by students. Have them discuss how this process is carried out in their local facility. Prior to the visit, encourage students to write down three questions that they would like to ask the expert. If an expert is not available to visit your classroom, organize a field trip for students to visit your local drinking water treatment plant and wastewater treatment plant.

 Testing Kit Extensions
While conducting both *Part I* and *Part II* of the activity, evaluate the effectiveness of each step by measuring water quality parameters such as turbidity, conductivity, pH, etc. using the Healthy Water, Healthy People testing kit. Be sure to establish a baseline by measuring these parameters before and after treatment.

 Internet Extension
Investigate the source for your municipal drinking water supply. Check for this information at the Web site: http://www.epa.gov/safewater/dwinfo.htm. Additionally, have students research and analyze the federal standards for allowable levels of drinking water and wastewater contaminants. Compare these standards with those of other countries if available.

Resources

American Chemical Society. 2002. *ChemCom: Chemistry in the Community*. New York, NY: W.H. Freeman and Company.

United States Environmental Protection Agency. 2001. *EPA Environmental Education*. Retrieved on February 4, 2002, from the Web site: http://www.epa.gov/safewater/kids/exper.html

The Watercourse. 1995. *Project WET Curriculum and Activity Guide*. Bozeman, MT: The Watercourse.

For information on drinking water treatment, visit the American Water Works Association Web site: www.awwa.org. For information on wastewater treatment, visit the Water Environment Federation Web site: www.wef.org.

Operator examining a floc settling tank. American Water Works Association

Drinking Water Worksheet

1. Cut your 2-liter bottle in half. Place about .5 liters of the source water in the bottom half. Observe the color and smell of the water and record your observations:

2. Aeration: To aerate, pour the water back and forth between the top and bottom halves ten times. End with the water in the bottom half of the bottle. Discuss and record any changes observed:

3. Coagulation: To remove suspended solids, alum is added to encourage coagulation and settling (flocculation) of the particles. Add a teaspoon of alum crystals to the source water. *(Important: Do not to touch the alum! If you do, immediately wash hands or affected areas with soap and water.)* Slowly stir the mixture for five minutes. You should see particles forming clumps (floc). Record changes in the water's appearance:

4. Sedimentation: Floc is allowed to settle out. Let the water rest undisturbed for 20 minutes (begin Step 5 while waiting). Record what is causing the floc to settle. _____ Record observations of the sample every five minutes:
    Start: _____ 5 mins: _____
    10 mins: _____ 15 mins: _____
    20 mins:_____

5. Filtration: Removes finer suspended particles. Construct a filter from the top half of your 2-liter bottle as follows:
    a. Remove the cap and place a coffee filter over the mouth, securing it with a rubber band. Turn the bottle upside down.
    b. Pour one cup (.24 l) of pebbles or aquarium gravel into the bottle. Pour one cup (.24 l) of coarse sand on top of the pebbles. Pour one cup (.24 l) of fine sand on top of the coarse sand.
    c. **Slowly** pour several cups of clean tap water through the filter to clean it. Try not to disturb the fine sand layer as you pour.
    d. Pour your source water through the sand filter while holding it over a large beaker.
    e. Compare your source water now to the source water you started with. Record observable changes:

Wastewater Worksheet

1. Assess the appearance and smell of your wastewater. Test the pH and check for the presence of oil by placing a drop on brown paper. If oil is present, an oil smear will form around the drop. If oil is not present, no smear will form.

2. Formulate a plan to clean your wastewater. Use any of the available cleaning products provided. Be sure to list a series of steps so that you can later evaluate the effectiveness of each step.

3. Evaluate the effectiveness of each step in your wastewater treatment plan. For example, was there an observable change as a result of each step? Record these changes as you implement your wastewater treatment plan.

4. Assess the appearance and smell of your wastewater. How does the wastewater compare to the observations made in question 1? Be sure to test the pH and for the presence of oil.

5. Were there some contaminants that were not removed? Could they be removed if you had other equipment or more time?

Drinking Water Treatment System

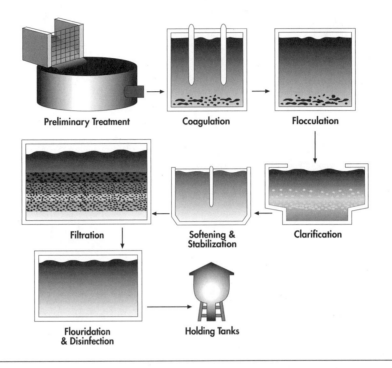

Preliminary Treatment   Coagulation   Flocculation

Filtration   Softening & Stabilization   Clarification

Flouridation & Disinfection   Holding Tanks

Wastewater Treatment

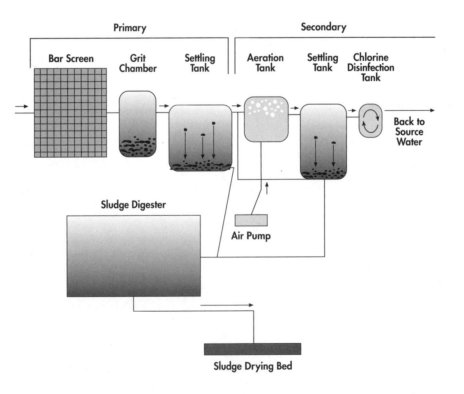

Primary   Secondary

Bar Screen   Grit Chamber   Settling Tank   Aeration Tank   Settling Tank   Chlorine Disinfection Tank

Back to Source Water

Sludge Digester

Air Pump

Sludge Drying Bed

PLANTS AND ANIMALS
ARE AFFECTED
BY CHANGES
IN WATER QUALITY

Benthic Bugs and Bioassessment

Summary

Students investigate the relative water quality of a stream by conducting a simulated bioassessment by sampling aquatic macroinvertebrates (represented by ordinary materials).

Objectives

Students will:

- investigate the role that aquatic macroinvertebrates play in determining water quality.
- simulate the process of rapid bioassessment of aquatic macroinvertebrates.
- collect, sort, classify, identify, analyze, and evaluate a sample of materials representing aquatic macroinvertebrates.
- determine a stream's water quality using a pollution tolerance index based on a sample of aquatic macroinvertebrates.
- compare the differences between the relative water quality of different samples.

Materials

- Copies of **Macroinvertebrate Identification Chart** Student Copy Page (1 per group)
- Copies of **Macroinvertebrate Data Sheets I** Student Copy Page (1 per group)
- Copies of **Macroinvertebrate Data Sheets II** Student Copy Page (1 per group)
- Copies of **Macroinvertebrate**

Grade Level:
6–12

Subject Areas:
Biology, Environmental Science, Mathematics, Language Arts

Duration:
Preparation: 30 minutes

Activity: two 50-minute periods

Setting:
Classroom

Skills:
Interpret, Organize, Gather, Communicate,

Vocabulary:
bioassessment, macroinvertebrate, biodiversity, benthic

Data Sheets III Student Copy Page (1 per group)
Materials for Bioassessment; similar objects may be substituted

- Plastic tubs or storage bins for holding samples–dishpan size (3)
- Smaller plastic tubs–white (3)
- Aquarium nets–small, hand-held; can use hands or other scoop (3)
- Ice cube trays, petri dishes, or other sorting devices (cups) (3)
- Calculators (3)
- Optional: Water (enough to fill the 3 sample tubs with at least 4 inches of water). (Optional: coloring for the water so students cannot see the objects in the sample tub. Coloring options include dark food coloring; powdered chocolate milk mix to simulate sediment, tea bags to darken the water, powdered fruit drink mix, other.)
- Small paper clips (100)
- Large paper clips (50)
- Six different sizes, shapes, or colors of beads (50 of each size/color/shape)
- Pennies or other coins (50)
- Thin rubber bands (50)
- Thick rubber bands (50)

Background

"The most direct and effective measure of the integrity of a water body is the status of its living systems" (Karr, 1998). One important way to determine the status of water's living systems is through biological assessment (bioassessment), which is the use of biological surveys and other direct measurements of living systems within a watershed. Aquatic macro-invertebrates (animals without backbones that live in aquatic environments and are large enough to be seen without the aid of a microscope or other magnification) are commonly monitored and are the basis of this activity.

Macroinvertebrates are valuable indicators of the health of aquatic environments in part because they are benthic, meaning they are typically found on the bottom of a stream or lake and do not move over large distances. Therefore, they cannot easily or quickly migrate away from pollution or environmental stress. Because different species of macroinvertebrates react differently to environmental stressors like pollution, sediment loading, and habitat changes, quantifying the diversity and density of different macroinvertebrates at a given site can create a picture of the environmental conditions of that body of water.

If exposed to an environmental stressor (e.g., pollution, warming due to low flows, low dissolved oxygen due to algal blooms, etc.), those macroinvertebrates that are intolerant to that stress may perish. Tolerant macroinvertebrates often inhabit the spaces left by the intolerant organisms, creating an entirely different population of organisms. For example, an unimpacted body of water will typically contain a majority of macroinvertebrates that are intolerant of environmental stressors, such

as mayflies (*Ephemeroptera*), stoneflies (*Plecoptera*), and Caddisflies (*Trichoptera*). A body of water that has undergone environmental stress may contain a majority of macroinvertebrates that are tolerant of these conditions such as leeches (*Hirudinea*), Tubifex worms (*Tubifex sp.*), and Pouch Snails (*Gastropoda*).

Bioassessments of macro-invertebrates are particularly helpful to biologists and others trying to determine the health of a river or stream. Bioassessment of macro-invertebrates is a procedure that uses inexpensive equipment, is scientifically valid if done correctly, and can be conducted by students. Bioassessments can provide bench-marks to which other waters may be compared and can also be used to define rehabilitation goals and to monitor trends. Trend monitoring is a common application of bioassessment by students groups and others involved in water quality monitoring.

Collecting, identifying, and quantify-ing macroinvertebrates are the initial steps in a bioassessment. The next step involves using formulas to calculate the relative water quality based on the diversity and quantity of the sampled organisms. These formulas, called metrics, relate the numerical diversity and density of organisms to a water quality rating. The most common metrics are the EPT/Midge Ratio and the Pollution Tolerance Index.

The EPT/Midge Ratio metric com-pares the total number of intolerant organisms, specifically the E.P.T—*Ephemeropterans* (mayflies), *Plecop-terans* (stoneflies), and *Trichopterans* (caddisflies)—with the total number of tolerant organisms, specifically *Chironomids* (midges). Typically the higher the number of intolerant organisms, the better the water quality.

The Pollution Tolerance Index assigns a numerical value to each macroinvertebrate order, with the higher numbers assigned to pollution intolerant organisms, and decreasing numbers assigned to increasingly pollution tolerant organisms. The scores are totaled and compared with a water quality assessment scale to yield a relative water quality rating for the sample.

To gather the best quality and most usable data, the Environmental Protection Agency (EPA) recom-mends that biological sampling of macroinvertebrates be conducted in ways that minimize year-to-year variability. To accomplish this, biologists tend to sample for at least one week during the same season(s) each year. Additionally, sampling is conducted when sites are easily accessible and the number of organ-isms is high. This usually occurs in the spring after the ice has broken and late-stage larvae are present, or in the late fall when organisms are more mature.

While bioassessments are extremely important in and of themselves, they are most useful when combined with chemical and habitat assessments. "Biosurvey techniques, such as the Rapid Bioassessment Protocols, are best used for detecting aquatic life impairments and assessing their relative severity. Once an impairment is detected, however, additional ecological data, such as chemical and biological (toxicity) testing is helpful to identify the causative agents, its source and to implement appropriate mitigation" (EPA, 1991).

Procedure

Warm Up

1.   Ask students to define the term "aquatic macroinvertebrate" (invertebrates that live in streams, rivers, lakes, or ponds that are large enough to be seen without the aid of a microscope or other magnification).

2.   Have them list examples of aquatic macroinvertebrates (e.g., leeches, mayflies, snails, dragonflies, etc.), and their role in the food web of a stream.

3.   Divide students into three groups and distribute copies of the ***Macroinvertebrate Identification Chart*** to each group. Instruct them to complete the middle (Looks Like) column of this sheet by researching aquatic macroinvertebrates on the Internet, in resource books, or in this guide (the activity "Invertebrates as Indicators" has invertebrate pictures). The North American Benthological Society maintains a Web site with links to various state and regional aquatic macro-invertebrates at www.benthos.org. Students will use this sheet in the bioassessment activity. *Option: To save time, you may complete the information in the chart for them.*

4.   Briefly explain to the students that aquatic macroinvertebrates are used as indicators of the relative health of a stream, and that the common form of sampling them is called a bioassessment, which they will conduct in this activity.

The Activity

1.   **Inform students that they will be simulating a bioassessment of a stream using ordinary objects to represent macroinvertebrates.**

2.   **Set up three sets of collecting stations (see illustration below), each containing the following:** stream sampling site (see directions in Step 2), collection bucket, sorting trays, the ***Macroinvertebrate Identification Chart,*** and ***Macroinvertebrate Data Sheets I, II, and III.***

Hellgrammites are used as water quality indicators. Courtesy: Richard Fields, Outdoor Indiana *Magazine*

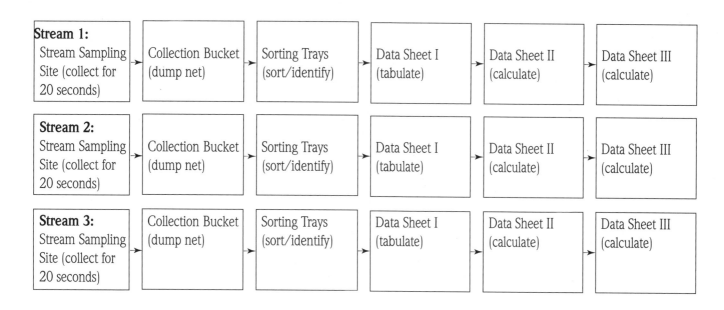

3. Optional: For the stream sampling sites: fill three large plastic storage bins with four inches of water and label them **Stream 1, 2, and 3.** *(Optional–add coloring to the water until objects on the bottom are not clearly seen).*
4. Place objects representing macroinvertebrates in the three tubs according to the following chart:

Macroinvertebrate	Represented by:	Number of Items per Sample			Total Items
		Stream Sample 1	Stream Sample 2	Stream Sample 3	
Mayflies	Yellow beads	35	15	0	50 beads
Stoneflies	Small paper clips	65	35	0	100 clips
Dobsonflies	Large paper clips	30	20	0	50 clips
Caddisflies	Red beads	30	20	0	50 beads
Craneflies	White beads	25	13	12	50 beads
Dragonflies	Green beads	20	20	10	50 beads
Scuds	Black beads	5	15	30	50 beads
Midges	Blue beads	0	20	30	50 beads
Leeches	Thick rubber bands	0	15	35	50 bands
Pouch Snails	Pennies	0	15	35	50 pennies
Tubifex Worms	Thin rubber bands	0	15	35	50 bands

5. Divide students into three groups. Assign students within each group to one of the following five tasks: stream sampling, sorting at the collection bucket, counting/recording at *Macroinvertebrate Data Sheet I,* and calculating/evaluating at *Macroinvertebrate Data Sheets II and III.*

6. Instruct students to simulate a rapid bioassessment at their stream sampling site as follows:

a. **Using an aquarium net, the students at the site have** *twenty seconds* **to collect as many macroinvertebrates (paper clips, beads, etc.) from the stream as possible.** They should place the macroinvertebrates in the collection bucket.

b. **Students at the collection bucket then sort the collected macroinvertebrates into like categories based on the *Macroinvertebrate Identification Sheet* and place them in the ice cube tray or cups.** For example, they should place all of the mayflies into one cube, caddisflies into another, etc.

c. **The students using the *Macroinvertebrate Data Sheet I* tabulate the sorting results onto the data sheet and calculate the percent composition of each macroinvertebrate in the stream site.**

d. **The students using *Macroinvertebrate Data Sheet II* take the data from *Macroinvertebrate Data Sheet I* to calculate the EPT/Midge ratio.**

Students conducting a bioassessment on a local stream. Courtesy: Richard Fields, Outdoor Indiana *Magazine*

e. The students with *Macroinvertebrate Data Sheet III* use the data from *Data Sheet I* to complete the Pollution Tolerance Index to determine their Water Quality Assessment score for their stream sample.

7. Have students compare their results with the other groups. What were the similarities and differences between the three sites? Which stream had the highest level of water quality? The lowest?

Wrap Up

Have students write a paragraph that describes their stream based on the macroinvertebrate sample they collected. If they sampled an impaired stream they should describe the habitat, address possible pollution sources, and give other pertinent details. Allow them to be creative.

Ask students what they think of this type of scientific sampling process. Do students feel that they could use this same process to perform a bioassessment in an actual stream? Did their samples accurately reflect the population of invertebrates in their stream? How do they know? Ask students to brainstorm how the process could be modified to increase its accuracy (e.g., conduct the sampling three times for each stream and compare or average the results)?

Have them identify positive and negative aspects of this type of

sampling. For example, do they believe that they netted larger insects more easily than smaller insects? Can such biased sampling occur in an actual rapid bioassessment of invertebrates?

Assessment

Have students:

- investigate the role that aquatic macroinvertebrates play in determining water quality (*Warm Up*).
- simulate the process of rapid bioassessment of aquatic macroinvertebrates (**Steps 5 and 6**).
- collect, sort, classify, identify, analyze, and evaluate a sample of materials representing aquatic macroinvertebrates (**Step 6**).
- determine a stream's water quality using a pollution tolerance index based on a sample of aquatic macroinvertebrates (**Step 6**).
- compare the differences between the relative water quality of different samples (**Step 7**).
- interpret water quality data to develop a description of a stream (*Wrap Up*).

Extensions

Have students conduct an actual rapid bioassessment of a local stream or lake. Ask a local biologist to assist in the assessment, or contact your local watershed monitoring group to see if students can help with a local monitoring project.

Have students locate specific aquatic macroinvertebrate identification keys for their watershed, state, or region.

Research the pollution tolerance, habitat, and regional distribution of the individual species. The North American Benthological Society maintains a Web site with links to various state and regional aquatic macroinvertebrates at www.benthos.org.

Resources

Fore, L. 1998. *Field Guide to Freshwater Invertebrates.* http://www.seanet.com/~leska/Online/about_guide.html.

Freshwater Benthic Ecology and Aquatic Entomology Homepage. 1999. http://www.chebucto.ns.ca/Science/SWCS/ZOOBENTH/BENTHOS/benthos.html

Karr, J., and E. Chu. 1998. *Restoring Life in Running Waters: Better Biological Monitoring.* Washington, D.C.: Island Press.

Mitchell, M., and W. Stapp. 1997. *Field Manual for Water Quality Monitoring: An Environmental Education Program for Schools.* Dubuque, IA: Kendall/Hunt Publishing Co.

Save Our Streams (SOS) Stream Study. 1999. http://www.people.virginia.edu/~sos-iwla/Stream-Study/StreamStudyHomePage/StreamStudy.HTML

United States Environmental Protection Agency (EPA). 1991. *Rapid Bioassessment Protocols (RBPs) for use in Streams and Wadeable Rivers.* http://www.epa.gov/owow/monitoring/rbp/

Macroinvertebrate Identification Chart

Macroinvertebrate	Looks like... (draw the invertebrate here)	Represented by... (for example beads, coins, etc)
Mayflies (Order *Ephemeroptera*)		
Stoneflies (Order *Plecoptera*)		
Caddisflies (Order *Trichoptera*)		
Dobsonflies (Order *Megaloptera*)		
Midges (Order *Chironomidae*)		
Craneflies (Order *Diptera*)		
Dragonflies (Order *Odonata*)		
Scuds (Order *Amphipoda*)		
Pouch Snails (*Class Gastropoda*)		
Tubifex Worms (Class *Oligochaeta*)		
Leeches (Class *Hirudinea*)		

Macroinvertebrate Data Sheet I

Stream #:

Recorded by:

Date of Sampling:

Percent Composition of Major Groups:

After the macroinvertebrates are sorted, tabulate the number of organisms for each of the major groups listed below and calculate their percent composition. This measure yields the relative abundance of macroinvertebrates within your sample.

Percent Composition = Number of Organisms in Each Group
    Total Number of Organisms

Macroinvertebrates	Number of Organisms in Each Group	Percent Composition
Mayflies (Order *Ephemeroptera*)		
Stoneflies (Order *Plecoptera*)		
Caddisflies (Order *Trichoptera*)		
Dobsonflies (Order *Megaloptera*)		
Midges (Order *Chironomidae*)		
Craneflies (Order *Diptera*)		
Dragonflies (Order *Odonata*)		
Scuds (Order *Amphipoda*)		
Pouch Snails (Class *Gastropoda*)		
Tubifex Worms (Class *Oligochaeta*)		
Leeches (Class *Hirudinea*)		
Total Number of Organisms		

(Adapted from Mitchell, 1997)

Macroinvertebrate Data Sheet II

Stream #: _____

Recorded by: _____

Date of Sampling: _____

EPT/Midge Ratio

1. Using the total number of macroinvertebrates from *Data Sheet I*, add up the total number of EPT individuals: Mayflies (***E****phemeroptera*) Total # _____

   Stoneflies (***P****lecoptera*) Total # _____

   Caddisflies (***T****richoptera*) Total # _____

   Total # (E+P+T) = _____

2. Add up the total number of Midges (*Chironomidae*)

   Total # Midges _____

3. Divide the total number of EPT individuals by the number of midges to determine the

   EPT/Midge Ratio = _____

4. Generally speaking, the larger the number of individuals in the EPT categories, the better the water quality. Therefore, the higher the final value of the ratio, the better the water quality.

EPT/Midge Ratio Formula:

$$\frac{\text{Total Number of EPT Individuals}}{\text{Total Number of Midge (\textit{Chironomidae}) Individuals}} = \underline{\qquad} =$$

Example 1: $\dfrac{\text{Total EPT}}{\text{Total Midges}} = \dfrac{100}{10} = 10$

Example 2: $\dfrac{\text{Total EPT}}{\text{Total Midges}} = \dfrac{40}{10} = 4$

Discussion:

The EPT/Midge Ratio of Example 1 is higher than Example 2; therefore the relative water quality is better in Example 1 than in Example 2.

Macroinvertebrate Data Sheet III

Pollution Tolerance Index

1. Place a check next to each macroinvertebrate group present in your sample. For example, whether you found one mayfly or fifty mayflies, place one check next to the mayfly line in Group 1.
2. Complete the chart for all of the macroinvertebrate groups.
3. Calculate the group scores using the multipliers provided.
4. Total all of the group scores for your Total Score.
5. Compare your Total Score with the Water Quality Assessment Chart scores and record the relative water quality rating for your stream sample.

Stream #:_____

Recorded by:_____

Date of Sampling:_____

Group 1 Macroinvertebrates: Very Intolerant	Group 2 Macroinvertebrates: Intolerant	Group 3 Macroinvertebrates: Tolerant	Group 4 Macroinvertebrates: Very Tolerant
____Stoneflies ____Mayflies ____Caddisflies ____Dobsonflies ____~~Dragonflies~~	____Dragonflies ____Scuds ____Craneflies	____Midges ____Leeches	____Pouch Snails ____Tubifex worms
# of checks = ____ x 4 Group Score=____	# of checks =____ x 3 Group Score=____	# of checks =____ x 2 Group Score=____	# of checks =____ x 1 Group Score=____

Total Score = _____ Your Water Quality Assessment:	Water Quality Assessment Chart: ≥23 Potentially Excellent Water Quality 17-22 Potentially Good Water Quality 11-16 Potentially Fair Water Quality ≤10 Potentially Poor Water Quality

(Adapted from Mitchell, 1997)

Water Quality Windows

Summary

Students explore the different water quality ranges required for the survival of aquatic and marine organisms by interpreting water quality data; sorting and classifying organisms according to their water quality requirements; and applying their knowledge to determine the effects of changes in water quality on organisms.

Objectives

Students will:

- explore the different water quality requirements of aquatic and marine organisms.
- determine organisms' water quality requirements and classify them accordingly.
- explore strategies for how organisms could survive acute changes in water quality.

Materials

- One copy of **Organism Cards** Teacher Copy Pages
- One copy of **Water Quality Windows** Teacher Copy Page
- One copy of **Answer Key** Teacher Copy Page
- Pencils and paper

Background

Every species has a habitat that is most favorable to its survival. For some, that habitat covers a diverse range of temperatures, terrain, and other conditions. For other species, the

Grade Level:
6–8

Subject Areas:
Ecology, Chemistry, Biology, Environmental Science

Duration:
Preparation: 30 minutes

Activity: two-50 minute periods (longer if Internet research is conducted)

Setting:
Classroom

Skills:
Match, Infer, Interpret, Sort and Classify, Research

Vocabulary:
pH, dissolved oxygen, extremophiles, estuary, marine, salinity

habitat requirements are very specific. Even humans, if deprived of our tools (clothing, fire, shelter), require a certain range of conditions for survival.

This is true of aquatic and marine organisms as well. Different species have adapted to the various water conditions and characteristics found throughout the world. Species survive in saltwater, freshwater and brackish (a mixture of both salt and fresh) conditions. Water quality factors like temperature, dissolved oxygen, salinity, and pH levels all affect the survival of aquatic and marine organisms.

Temperature: Temperature influences several other parameters in a water body. For example, cold water is able to hold more dissolved oxygen than warm water. As the water temperature rises, dissolved oxygen is forced out of solution, thereby lowering dissolved oxygen levels. Aquatic organisms' metabolic rates also increase in warm water, demanding even more dissolved oxygen. Temperature is measured in Celsius in this activity.

The ability of organisms to fight disease, parasites, or pollution is compromised when water temperature increases and available dissolved oxygen decreases. Eventually if cool waters are warmed for an extended period of time, cool water species such as trout and stonefly nymphs may be replaced with warm water species such as carp and dragonfly nymphs. In warm waters, the fish and aquatic macroinvertebrates that live there are adapted to survive in warmer waters and lower levels of dissolved oxygen. If temperatures increase to where dissolved oxygen levels are greatly decreased, the organisms in that water may have to adapt, move on, or perish.

Dissolved Oxygen (DO): Most aquatic plants and animals, like humans, need oxygen to live. Oxygen is dissolved in water in varying amounts, depending on many factors. Swift moving, well-aerated streams have a high level of dissolved oxygen, whereas stagnant ponds have a very low level of dissolved oxygen. Cool water holds onto dissolved oxygen longer than warm water does—dissolved oxygen is forced out of solution as water temperatures rise. In general, higher dissolved oxygen levels indicate healthier water for aquatic organisms. Dissolved oxygen is typically measured in mg/L. Dissolved oxygen levels <4 - 5 mg/L can become fatal to fish if they cannot migrate to a place with higher dissolved oxygen levels.

Salinity: Salinity is the measure of the dissolved salts in water. The average salinity of ocean water is 35 g/L, meaning there are 35 grams of dissolved ions per 1000 grams of sea water. This can also be stated as 35 parts per thousand, or ppt. Chloride and sodium make up 90% of the dissolved content in saline waters. The highest levels of salinity are found in the center of the

oceans where the input of rivers and streams does not dilute the salinity, and in subtropical regions where there are large amounts of evaporation.

The salts in the oceans come from the weathering of continents, deep sea hydrothermal vents, and submarine volcanoes. The salinity range of the oceans varies with the seasons and location due to the melting of ice, inflow of river water, evaporation, rain, snowfall, wind, wave motion, and ocean currents.

pH: Because aquatic organisms adapt to water of a certain pH range, any fluctuation in the pH of their environment can lead to stress or death. Water (H_2O) is made up of hydrogen (H^+) and hydroxide (OH^-) ions. The H^+ concentration is used to quantitatively describe the acidity of a liquid.

Often these concentrations are quite small, so the negative logarithm of the hydrogen concentration has been converted to the easy-to-use pH scale. In this scale, a pH of seven is neutral, with equal numbers of hydrogen and hydroxide ions. A pH above seven is basic, with more OH^- ions than H^+ ions. Conversely, a pH below seven is acidic, containing more H^+ ions than OH^- ions. Since the scale is logarithmic, for every one unit of change on the scale there is a tenfold change in the acidity of the sample. For example, a river with a pH of 5 is 10 times more acidic than a river with a pH

of 6, and 100 times more acidic than a river with a pH of 7.

There are several factors that affect the pH of natural waters, some natural, and some human induced. The geology of a riverbed and surrounding rocks often dictates the pH of the river that flows through it. Limestone is basic; therefore rivers that run through limestone tend to have a pH above or around seven. Human influences such as the burning of fossil fuels can cause rain, snow, and rivers to be more acidic (called acid rain).

Whether in the near-frozen polar oceans or the scalding hydrothermal pools in Yellowstone National Park, aquatic organisms are adapted to survive within specific ranges of water conditions. Generally most species live within a temperature range of 0°C to 40°C and in waters with a pH of between 6 and 9. However, some microbes, known as extremophiles, have been found in water temperatures up to 113°C and pH levels as low as 2.

Because many species are adapted to specific conditions, they are very vulnerable to change. Weather patterns, land use changes, and pollution can change water quality enough to significantly affect habitat. In circumstances of dramatic change, aquatic organisms must find protection from these changes or migrate to more suitable water. Understanding the unique water quality requirements of aquatic organisms can help us become better stewards of our watersheds for the organisms that live there.

Procedure
Warm Up
Aquatic (freshwater), marine (saltwater), and estuarine (mixture of saltwater and freshwater) organisms have *Water Quality Windows* within which they live. Ask students to name an aquatic, marine, or estuarine organism and a general water quality window within which

that organism lives (e.g., a frog – must live in or near water; trout – need cold, clean water to thrive; goldfish – can live in warm or cool water; sharks – most need highly salty, open ocean waters to thrive; etc.) Explain to students that just as humans have specific requirements for survival, so do aquatic and marine organisms. This activity will explore the water quality requirements of various aquatic organisms.

Review with students the basic concepts of temperature, dissolved

oxygen, salinity, and pH. Explain the units of measurement for each, as well as their relevance to aquatic and marine organisms.

The Activity
Part I
1. Cut out the *Water Quality Windows* cards. Tape or hang them around the room in numbered order, allowing room for students to move to each card.
2. Cut out the *Organism Cards* and distribute one to each

Thermophilic bacteria, or extremophiles, can live in geothermal areas such as this boiling hot cauldron in Yellowstone National Park. Courtesy: Yellowstone National Park; National Park Service

student. Remind students that organisms that live in aquatic or marine environments often have a "window" of water quality within which they live. These parameters help form this "window."

3. Ask students to study their *Organism Card.* Ask a student who has a Freshwater card to read the parameters on their card. Ask a student with a Saltwater card to read their parameters. Why are the parameters different between the freshwater and saltwater cards? Is salinity typically a condition of freshwater?

 Option: Have students research their organism in depth. Students could investigate and describe their organism in detail, the ranges for other water quality parameters, where the organism is found, and other relevant facts. See the **Resources** section for references.

4. Instruct students to investigate the water quality windows that are placed around the room. Explain that their goal is to find the window that their organism fits into best, and when they find it, to stand next to that *Water Quality Window* **card.** *All* of their parameter ranges *must* fit within the ranges on the *Water Quality Window* card to be the *best* fit. This process may take some time as the students begin to realize that they must look carefully at all of their parameters to find the best fit. Refer to the **Answer Key** *Teacher Copy Page* for clarification.

5. There is one fish, the Spotted Seatrout *(Cynoscion nebulosus),* **that appears twice as it can fit into two of the Water Quality Windows.** The two students with these cards may choose the same or different windows.

Part II

1. The students should now be sorted into groups based on their organism's water quality requirements. **Have them note what other organisms fit into their** *Water Quality Window.* Ask students if there are any organisms in their window that they didn't expect to see?

2. Ask the groups of students to work together to complete the following:
   a. Make a list of the organisms found in their *Water Quality Window.*
   b. Brainstorm what the habitat of their *Water Quality Window* might look like. Have them consider:
       i. whether it is freshwater or saltwater;
       ii. where in a watershed, ocean, or other location it may be found—high in the mountains or near the mouth of a river; deep ocean or along the shore line, etc.
       iii. whether it could be a lake, stream, estuary, open ocean, or other location;
       iv. how the organisms within a window might interact—do any of them feed on each other, depend on similar foods, etc.

   c. **Option:** *If time allows, have students draw their organisms and what their habitat might look like based on the questions above.*
   d. Have the students present their findings to the class.

3. With the students gathered at their *Water Quality Windows,* **read the following scenario aloud:**

A severe drought has hit the entire region that the organisms live in, from the highest mountains to the open oceans. With the drought, the temperatures have hit record highs, while the precipitation levels are at record lows.

The water temperatures of the freshwater rivers and lakes in the region have risen dramatically. For over two weeks, the freshwater temperatures have consistently stayed at the highest temperature of each Water Quality Window. For example, the water temperature for Water Quality Window #1 is now 22C, and so on for the rest of the freshwater windows.

Without any rain, the salinity level in the saltwater estuaries and oceans has risen as well. The salinity is now consistently staying at the highest salinity reading of each Water Quality Window. For example, the salinity for Water Quality Window #6 is now 34 ppt, and so on for the rest of the saltwater windows.

4. Have the students compare their organism's require-

ments with the new temperature and salinity readings produced by the drought. Can they still fit within their original *Water Quality Window* based on the new readings? (For example, the Redbreast Sunfish can stay in its original *Water Quality Window* with the temperature at 22C, as it is still within its range. However, the Grayling's temperature range is 6-18C, so it cannot still fit within its original *Water Quality Window.*)

5. Have students who cannot stay within their original *Water Quality Window* raise their hands. Ask them how they plan on surviving the drought based on their water quality needs? What strategies could they use to survive? (Possibly moving to a different window, searching for deeper and colder water, moving to less saline waters, etc.)

6. Instruct the students who can no longer stay within their *Water Quality Windows* to move to a different window if they can. Ask the students what impacts, if any, could these newcomers have on the organisms that currently live in those habitats? (May be a predator of those organisms.) What could happen to those organisms that cannot fit within *any* window due to the drought?

Part III

1. Before you begin, ask students to explain the relationship between temperature and dissolved oxygen. Provide background as needed—see **Background** section.

2. Repeat Part II of the activity,

only this time explain to students that they are to use dissolved oxygen as the freshwater parameter that is impacted.

3. Instruct them to consider the same drought scenario as described in Part II, only this time the new readings on the freshwater *Water Quality Window* cards show the dissolved oxygen levels are at the lowest point in their ranges. For example, the dissolved oxygen level for Water Quality Window #1 is now 8.5 mg/L, and so on for the other *Water Quality Windows.*

4. The saltwater *Water Quality Window* cards should stay the same as the scenario from Part II.

3. Repeat Steps 5 and 6 from Part II.

Wrap Up

Ask how many of the students survived the drought when the temperatures changed? How many of them would be in peril if the temperatures and salinities kept rising? How many students survived the drought when the dissolved oxygen levels changed? In reality, the temperature and dissolved oxygen levels are closely linked. Ask students to explain this relationship, or have them research it on their own (www.healthywater.org or the *Healthy Water, Healthy People Testing Kit Manual* are great resources for this information).

Have students brainstorm ways that aquatic and marine organisms

actually do survive acute changes in water quality (move to a better habitat, slow their metabolism down by becoming less active, seeking out springs, deep holes, or riffles with colder temperatures and/or higher dissolved oxygen levels).

Have students write a paragraph detailing the water quality window of an aquatic or marine organism of their choice. They can use one of the data cards or conduct further research on their own (see Internet Extensions and Resources sections for additional resources).

Assessment
Have students:
- explore the different water quality requirements of numerous aquatic and marine organisms (*Warm Up, Parts I, II, III*).
- determine organisms' water quality requirements and classify them accordingly (*Parts I, II, III*).
- explore strategies for how organisms could survive acute changes in water quality (*Wrap Up*).

Extensions
Challenge students to find the best fit(s) for the following fish. These are tricky in that they can inhabit more than one water quality window. How likely is a carp to adapt to changing environmental conditions? How about a salmon?

Common Carp (*Cyprinus carpio*)
(freshwater, sometimes brackish
water)
Temperature: 3 – 32°C
D. 13.5 – 7.3 mg/L
pH: 7.0 – 9.0

Sockeye Salmon (*Onchorynchus
nerka*) (lives in saltwater, spawns in
freshwater)
Temperature: 0 – 25°C
D. 14.6 – 8.3 mg/L
pH: 6.0 – 8.0

Atlantic Salmon (Salmo salar) (*lives
in saltwater, spawns in freshwater*)
Temp: 2-22° C
DO: 13.8 – 8.7 mg/L
pH: 6.0 – 8.0

Have students research the temperature, pH ranges, and salinity of local
waters, and use that information to
make educated guesses about what
type of fish could be found in that
water. Check hypotheses with your
local fish and wildlife agency or
biologist.

Invite a fisheries or aquatic biologist
to your classroom or on a field trip to
discuss the diversity of aquatic
species living in nearby waters.
Instruct your students to brainstorm
and write up three questions to ask

the expert about water quality as
related to organisms.

 Internet Extensions:
Have students find a
picture and other information about
their organism. Other information
may include related species, where
the organism lives (countries), other
water quality requirements (e.g.,
spawning temperatures, etc.), and
additional references to research.
There are several useful Web sites in
the **Resources** section.

Testing Kit Extensions
Have students test local
waters for temperature, pH and
dissolved oxygen concentrations
using Healthy Water, Healthy People
testing kits. What fish would they
expect to find within your local
window of water quality? Ground-
truth the students' predictions by
inviting a fisheries biologist to
accompany the students in the field.
If possible, have the biologist sample
the area to determine what fish
species live there.

Resources

Froese, R. and D. Pauly, ed. 2002.
FishBase. Retrieved on March 4,
2002, from the Web site: http://
www.fishbase.org

Funderburk, S., S. Jordan, J.
Mihursky, and D. Riley, eds. 1991.
*Habitat Requirements For Chesa-
peake Bay Living Resources;* Second
Edition. Annapolis, MD.: Chesa-
peake Bay Program.

Piper, R., I. McElwain, L. Orme, J.
McCraren, L. Fowler, and J.
Leonard. 1982. *Fish Hatchery
Management.* Washington, D.C.:
United States Department of the
Interior.

Revenaugh, J. 2002. *Geologic
Principles, Lecture 11 Oceans and
Coasts.* Retrieved in January 2003
from the Web site: http://
www.es.ucsc.edu/%7Ees10jsr/
classnotes/Lectures/lecture.11.html

Royer, T. and T. Weingartner. *Gak1
Time Series.* Retrieved in January
2003 from the Web site: http://
halibut.ims.uaf.edu:8000/gak1/

Wetzel, R. 2001. *Limnology.* San
Diego, CA.: Academic Press.

Wills, L. 2000. *Species Information System.* Retrieved in January
2003, from the Web site: http://
fwie.fw.vt.edu/WWW/macsis/
fish.htm

Notes:

Organism Cards

Pacific Halibut *Hippoglossus stenolepis* Temperature: 3-8°C Salinity: 27.5-33.5 ppt **Saltwater**	**Northern pike** *Esox lucius* Temperature: 10-28°C dissolved oxygen: 7.8-11.3 mg/L pH: 5-9 **Freshwater**	**Flathead catfish** *Pylodictis olivaris* Temperature: 20-32.5°C dissolved oxygen: 7.3-9.1 mg/L pH: 7-9 **Freshwater**
Pacific cod *Gadus macrocephalus* Temperature: 7-8°C Salinity: 12.7-23 ppt **Saltwater**	**Rainbow trout** *Oncorhynchus mykiss* Temperature: 10-24°C dissolved oxygen: 8.4-11.3 mg/L pH: 6.7-8 **Freshwater**	**Microbe** *Sulfolobus acidocaldarins* Temperature: 60-90°C dissolved oxygen: 2.8-4.7 mg/L pH: 2-3 **Hot Springs**
Cutthroat trout *Oncorhynchus clarki clarki* Temperature: 12.7-20°C dissolved oxygen: greater than 6 mg/L pH: 4.5-9 **Freshwater**	**Largemouth bass** *Micropterus salmoides* Temperature: 10-32°C dissolved oxygen:: 7.3-11.3 mg/L pH: 7-9 **Freshwater**	**Microbe** *Pyrolobus fumarii* Temperature: 90-113°C dissolved oxygen: 1.6-2.8mg/L pH: 4-6.5 **Deep Sea Hydrothermal Vents**
Spotted Seatrout *Cynoscion nebulosus* Temperature: 15-27°C Salinity: varies **Saltwater**	**Rock bass** *Ambloplites rupestris* Temperature: 10-30°C dissolved oxygen: 7.6-11.3 mg/L pH: 7.0-9.0 **Freshwater**	**Microbe** *Polaromanas vacuolata* Temperature: 0-10°C dissolved oxygen: 11.3-14.6 mg/L pH: 6-8 **Antarctic Sea Ice**
Brown Trout *Salmo trutta trutta* Temperature: 10-24°C dissolved oxygen: 8.4-11.3 mg/L pH: 6.7-8 **Freshwater**	**Channel catfish** *Ictalurus punctatus* Temperature: 10-32°C dissolved oxygen: 7.0-11.3 mg/L pH: 7.0-9.0 **Freshwater**	**Grayling** *Prototroctes oxyrhynchus* Temperature: 6-18°C pH: 7-7.5 dissolved oxygen: Varies **Freshwater**

Organism Cards

Dungeness crab *Caner magister* Temperature: 8-10°C Salinity: varies **Saltwater**	**Blue marlin** *Makaira nigican* Temperature: 21-27°C Salinity: 35-36 ppt **Saltwater**	**Slimy sculpin** *Cottus cognatus* Temperature: 4-16°C dissolved oxygen: 9.9-13.1 mg/l pH: 6-8 **Freshwater**
Blue crab *Callinectes sapidus* Temperature: 1-34°C Salinity: 3-15 ppt **Saltwater**	**Black sea bass** *Centropristis striata* Temperature: 6-29°C Salinity: 1-36 ppt **Saltwater**	**Redbreast sunfish** *Lepomis auritus* Temperature: 4-22°C dissolved oxygen: 8.7-13.1 mg/l pH: 7.0-7.5 **Freshwater**
White perch *Morone americana* Temperature: 2-32.5°C Salinity: 5-18 ppt **Saltwater**	**Spiny lobster** *Panulirus argus* Temperature: around 20°C Salinity: not below 19 ppt **Saltwater**	**Brook trout** *Salvelinus fontinalis* Temperature: 6-20°C dissolved oxygen: 9.1-12.4 mg/l pH: 6-8 **Freshwater**
Spotted seatrout *Cynoscion nebulosus* Temperature: 15-27°C Salinity: varies **Saltwater**	**American oyster** *Crassostrea virginica* Temperature: 20-30°C Salinity: 20-35 ppt **Saltwater**	**Arctic char** *Salvelinus alpinus* Temperature: 4-16°C dissolved oxygen: unknown **Freshwater**

Answer Key

Water Quality Window #1
Freshwater
Temp: 4-22°C
DO: 8.5-13.5 mg/l
pH: 6-8
- Redbreast sunfish
- Slimy sculpin
- Brook trout
- Grayling
- Artic Char

Water Quality Window #2
Freshwater
Temp: 10-28°C
DO: 6.5-11.3 mg/l
pH: 5-9
- Brown trout
- Cutthroat trout
- Rainbow trout
- Northern pike

Water Quality Window #3
Freshwater
Temp: 10-32°C
DO: 7-12 mg/l
pH: 7-9
- Largemouth bass
- Rock bass
- Channel catfish
- Flathead catfish

Water Quality Window #4
Saltwater
Temp: 1-34°C
Salinity: 3-18 ppt
- Blue crab
- White perch
- Spotted seatrout

Water Quality Window #5
Saltwater
Temp: 6-30°C
Salinity: 1-36 ppt
- Blue marlin
- Black sea bass
- Spiny lobster
- American oyster
- Spotted seatrout

Water Quality Window #6
Saltwater
Temp: 3-10°C
Salinity: 7-34 ppt
- Pacific cod
- Dungeness crab
- Pacific Halibut

Water Quality Window #7
Extreme Environments

Microbe 1: (*Sulfolobus acidocaldarins*)
Hot Springs; Temp: 60-90°C DO: 2.8-4.7 pH: 2-3

Microbe 2: (*Pyrolobus fumarii)*
Deep Sea Hydrothermal Vents
Temp: 90-113°C DO: 2.8-1.6mg/L pH: 4-6.5

Microbe 3: (*Polaromanas vacuolata*) **Antarctic Sea Ice**
Temp: 0-10°C DO: 11.3-14.6mg/L pH: 6-8

Water Quality Windows

Water Quality Window **#1** **Freshwater** Temperature: 4-22°C DO: 8.5-13.5 mg/L pH: 6-8	Water Quality Window **#5** **Saltwater** Temperature: 6-30°C Salinity: 1-36 ppt
Water Quality Window **#2** **Freshwater** Temperature: 10-28°C DO: 6.5-11.3 mg/L pH: 5-9	Water Quality Window **#6** **Saltwater** Temperature: 3-10°C Salinity: 7-34 ppt
Water Quality Window **#3** **Freshwater** Temperature: 10-32°C DO: 7-12 mg/L pH: 7-9	Water Quality Window **#7** **Extreme Environments** (organism does not fit within other windows)
Water Quality Window **#4** **Saltwater** Temperature: 1-34°C Salinity: 3-18 ppt	

Invertebrates as Indicators

Summary

Students simulate the effects of environmental stressors on aquatic macroinvertebrate populations, record changes to these populations, and investigate how impacted populations recover.

Objectives

Students will:

- illustrate how tolerance to water quality conditions varies among macroinvertebrate organisms.
- explain how population diversity provides insight about the health of an ecosystem.
- determine how to improve the water quality of a stream and therefore the diversity of macroinvertebrates by reducing pollution.

Materials

- **Macroinvertebrate Illustrations**
  Teacher Copy Page (one illustration per student) (Divide the number of students by 7 and make that number of copies of each macroinvertebrate picture. One side of each label should have a picture of one of the seven macroinvertebrates. The other side of each label (except for midge larvae and rat-tailed maggots) should have a picture of either the midge larva or rat-tailed maggot. For durability, the cards may be laminated. Use clothespins or paper

Grade Level:
6–8

Subject Areas:
Ecology, Environmental Science, Mathematics

Duration:
Preparation:
Part I:
20 minutes
Part II:
50 minutes

Activity:
Part I:
50 minutes
Part II:
50 minutes

Setting:
Large playing field or gymnasium

Skills:
Interpreting, Analyzing, Predicting, Cause and Effect

Vocabulary:
macroinvertebrate, biodiversity, bioassessment, indicator

clips to attach labels to students' clothing.

- *Pillowcases or burlap sacks*
- *Copies of **Data Chart** Teacher Copy Page* (2 copies)
- *Samples of macroinvertebrate organisms* (optional)
- *Hula hoops, rope, or cloth strips*

Background

Macroinvertebrates (organisms that lack an internal skeleton and are large enough to be seen with the naked eye) are an integral part of wetland and stream ecosystems. Examples of macroinvertebrates include mayflies, stoneflies, dragonflies, rat-tailed maggots, scuds, snails, and leeches. These organisms may spend all or part of their lives in water. Usually their immature phases as larvae or nymphs (maggot is the term for the larva of some flies) are spent entirely in water. Larvae do not show wing buds and usually look very different than the adult insects. Nymphs generally resemble adults, but have no developed wings and are usually smaller.

Because different species of macroinvertebrates respond differently to a variety of environmental stress factors, their presence and diversity in a body of water can often give a good indication of the general health of that water. Thus, water quality researchers often sample macroinvertebrate populations to monitor changes in stream conditions over time and to assess the cumulative effects of environmental stressors. Healthy streams are usually characterized by the presence of a diversity of aquatic organisms along with a variety of food sources, adequate oxygen levels, and temperature ranges conducive to their survival. When environmental degradation upsets this balance, diversity of an aquatic community often decreases as less tolerant organisms are eliminated, the more tolerant ones survive, and truly tolerant ones thrive.

A variety of environmental stressors, including urban and agricultural runoff, sewage and fertilizers, changes in land use, destruction of waterside vegetation, stream channelization and the introduction of alien species affect macro-invertebrate populations. These events and activities can lead to various forms of pollution including: sediment and nutrient loading from runoff, increased erosion from land use changes (i.e., construction, poorly managed cropland, etc.), increased water temperatures from destruction of waterside vegetation or sedimentation.

Depending on the species, macroinvertebrates respond differently to various environmental stressors. In the case of intolerant species like mayfly and stonefly nymphs, and caddisfly larvae, pollution-caused changes in stream conditions have serious effects. Some leave to find more favorable habitats. Others stop reproducing or, if the environmental stress is severe enough, die altogether. Other more

tolerant species (facultative organisms) like dragonfly and damselfly nymphs prefer good stream quality but can survive in polluted conditions. Truly tolerant organisms like rat-tailed maggots and midge larvae will thrive in polluted conditions.

Generally then, a sample of macro-invertebrates consisting of rat-tailed maggots, snails, and dragonfly nymphs is a strong indicator of poor water-quality conditions like low oxygen levels, increased sediment and the presence of contaminants. On the other hand, a sample containing a diversity of organisms is a likely indicator that stream health is good. Baseline data is essential, however, because some healthy streams may naturally contain only a few macroinvertebrate species.

Procedure

Warm Up

Review the conditions that are necessary for a healthy ecosystem. Ask students to describe what could happen to an ecosystem if these conditions are altered or eliminated. What clues might students look for to determine the health of an ecosystem?

Remind students that a stream is a type of ecosystem. Ask them how they would assess the health of a stream. Suggest that students could conduct a visual survey of the surrounding area and answer the following questions: What land use practices are visible? How might these practices affect the stream? Is there plant cover on the banks of the

stream or are the banks eroded? What color is the water? What is living in the stream?

Identify several environmental stressors (e.g., urban and agricultural runoff, sedimentation, low dissolved oxygen) and discuss how they can affect the health of a stream. Review the many types of plants and animals, including insects, which live in streams. How might environmental stressors affect these organisms? Would all organisms be impacted in the same way? Why or why not?

The Activity

Note: Adapt this activity for your area by using local species. Contact your local water monitoring group or state department of environmental quality or fisheries.

Part I

1. **Tell students they are going to play a game that simulates changes in a stream when an environmental stressor, such as a pollutant, is introduced.** Show students the playing field and indicate the boundaries.
2. **Choose one student volunteer to be an environmental stressor and let the other students decide the type of stressor (e.g., sedimentation, sewage, or fertilizer).** Discuss ways that a stream can become polluted and how this can alter stream conditions. (If the class is large or the playing field is big, more students will need to be stressors.)

Students observing aquatic macronvertebrates collected from a local stream. Courtesy: Richard Fields, Outdoor Indiana Magazine

3. **Divide the rest of the class into seven groups. To play the game, each group represents one type of macroinvertebrate species listed in *Macroinvertebrate Groups*.** Record the number of members in each group on the ***Data Chart*** *Teacher Copy Page.* (Note: Try to have at least four students in each group. For smaller classes, reduce the number of groups to keep the group size small. For example, eliminate the stonefly nymph and the damselfly nymph groups.)

4. **Distribute appropriate identification labels to all group members.** The picture of each group's macroinvertebrate should face outward when labels are attached.

5. **Inform students that some macroinvertebrates have hindrances that will affect their ability to cross the field. Tell each species what their hindrance will be** *(See Intolerant Macroinvertebrates and Hindrances.)* These obstacles symbolize sensitive organisms' intolerance to pollutants. Have students practice their particular motions/hindrances.

6. **Assemble the macroinvertebrate groups at one end of the playing field and the environmental stressor(s) at midfield.** When a round starts, macroinvertebrates will move toward the opposite end of the field and the stressor will try to tag them. To "survive," the macroinvertebrates must reach

the opposite end of the field without being tagged by the environmental stressor. The environmental stressor can try to tag any of the macro invertebrates.

7. **Begin the first round of the game.** Tagged macroinvertebrates must go to the sideline and flip their identification labels to display the more tolerant species (e.g., rat-tailed maggot or midge larva). Tagged players who are already in a tolerant species group do not flip their labels.

8. **The round ends when all of the macroinvertebrates have either been tagged or have reached the opposite end of the playing field.** Record the revised number of members in each species on the **Data Chart** *Teacher Copy Page.*

9. **Complete two more rounds, with all tagged players rejoining the macroinvertebrates who successfully survived the previous round.** Record the number of members in each species of macroinvertebrates at the conclusion of each round. Because some players will have flipped their identification labels, there will be a larger number of tolerant species in each successive round.

Part II

1. **Gather students and have them briefly describe what happened to the** macroinvertebrate populations in *Part I.*

2. **Remind them of their chosen environmental stressor(s), and ask them to brainstorm methods for reducing or eliminating these stressors from a stream environment.** For example, if the stressor was sediment, ways to reduce it include leaving vegetation buffer strips along streams, creating barriers around construction and other changed land use sites, etc. If the stressor was acid rain, one preventive method is using less fossil fuel by carpooling or biking to school, installing scrubbers in industrial smoke stacks, etc. *(Option: A break to research prevention methods is appropriate, or students could research them as homework.)*

3. **Building on this discussion, ask students to imagine implementing their chosen stressor-reduction methods, thereby improving the water quality of the stream in which they (as macroinvertebrates) live.** To represent the improved water quality and macroinvertebrate habitat, randomly toss hula-hoops or other flat-lying indicators (ropes, cloth strips) into the playing field.

4. **These areas will represent improved habitat and water quality.** Play the game again, allowing the macroinvertebrates to use the safety zones (where they cannot be tagged) to elude the stressor(s).

5. **Play three rounds using these safety zones and record the results on a second copy of the Data Chart** *Teacher Copy Page.* Compare the results of the macroinvertebrate populations in *Part I* with *Part II.*

Wrap Up and Action

Discuss the outcomes of *Parts I* and *II.* Emphasize the changes in the distribution of organisms among groups. Have students compare population sizes of groups at the beginning and end of the game and provide reasons for the changes. Review why some organisms tolerate poor environmental conditions and others don't. Have students compare the stream environment at the beginning of the game, again after *Part I,* and then after *Part II.*

Challenge students to apply the concepts from the activity by having them research or actually collect a sample of aquatic macroinvertebrates from a nearby stream (local watershed monitoring groups may have data sets). If this data is unavailable, create a mock data set for a healthy stream and an impaired stream to share with your students. Have them identify the relative health of the streams based on the data, or even speculate where the sample was taken (headwaters, below a city, etc.).

Assessment

Have students:

- illustrate how tolerance to water quality conditions varies among macroinvertebrate organisms (**Warm Up, Part I**).
- explain how population diversity provides insight into the health of an ecosystem (**Part I**).
- determine how to improve the water quality of a stream and therefore the diversity of macroinvertebrates by reducing pollution (**Part II**).

Extensions

Introduce the practice of sampling macroinvertebrate populations to monitor stream quality. Review the methods of bioassessment in the activity "Benthic Bugs and Bioassessment" page 154 for an overview of how to sample for macroinvertebrates. This activity also includes methods for identifying, sorting, classifying, and analyzing data using metrics. The end result will be a water quality rating based on the macroinvertebrates present in the stream.

Testing Kit Extensions

Design and conduct a stream-monitoring project on a local waterway using Healthy Water, Healthy People Testing Kits, specifically the Healthy Water, Healthy People MacroPac™ Macroinvertebrate Sampling Kit.

Resources

Ancona, G. 1990. *River Keeper*. New York, N.Y.: Macmillan.

Edelstein, K. 1993. *Pond and Stream Safari: A Guide to the Ecology of Aquatic Invertebrates*. Ithaca, N.Y.: Cornell University.

Fore, L. 1998. *Field Guide to Freshwater Invertebrates*. http://www.seanet.com/~leska/Online/about_guide.html

Freshwater Benthic Ecology and Aquatic Entomology Homepage. 1999. http://www.chebucto.ns.ca/Science/SWCS/ZOOBENTH/BENTHOS/benthos.html

Karr, J., and E. Chu. 1998. *Restoring Life in Running Waters: Better Biological Monitoring*. Washington, D.C.: Island Press.

Mitchell, M., and W. Stapp. 1997. *Field Manual for Water Quality Monitoring: An Environmental Education Program for Schools*. Dubuque, IA.: Kendall/Hunt Publishing Co.

Save-Our Streams (SOS) Stream Study. 1999. http://www.people.virginia.edu/~sos-iwla/Stream-Study/StreamStudyHomePage/StreamStudy.HTML

The Stream Scene: Watersheds, Wildlife and People. 1990. Portland, OR: Oregon Department of Fish & Wildlife.

United States Environmental Protection Agency (EPA). 1991. *Rapid Bioassessment Protocols (RBPs) for use in Streams and Wadeable Rivers*. http://www.epa/gov/owow/monitoring /rbp/

Notes:

Macroinvertebrate Groups
Caddisfly larva
Mayfly nymph
Stonefly nymph
Dragonfly nymph
Damselfly nymph
Midge larva
Rat-tailed maggot

Intolerant Macroinvertebrates and Hindrances		
ORGANISM	HINDRANCE	RATIONALE FOR HINDRANCE
Caddisfly	Must place both feet in a sack and hop across field, stopping to gasp for breath every five hops.	Caddisflies are intolerant of low oxygen levels.
Stonefly	Must do a push-up every ten steps.	When oxygen levels drop, stoneflies undulate their abdomens to increase the flow of water over their bodies.
Mayfly	Must flap arms and spin in circles when crossing field.	Mayflies often increase oxygen absorption by moving gills.

Data Chart

ORGANISM	TOLERANCE	START	ROUND 1	ROUND 2	ROUND 3
Caddisfly larva	Intolerant				
Mayfly nymph	Intolerant				
Stonefly nymph	Intolerant				
Dragonfly nymph	Facultative				
Damselfly nymph	Facultative				
Midge larva	Tolerant				
Rat-tailed Maggot	Tolerant				
TOTAL					

Macroinvertebrate Illustrations

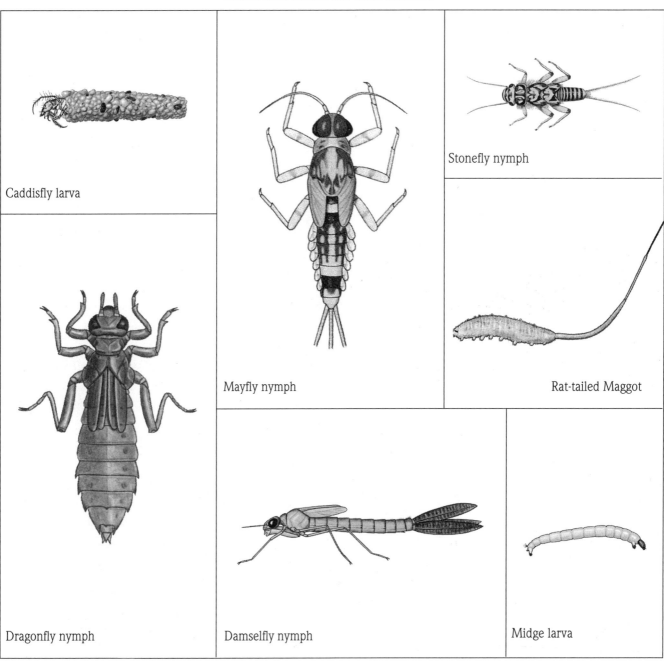

Caddisfly larva

Mayfly nymph

Stonefly nymph

Rat-tailed Maggot

Dragonfly nymph

Damselfly nymph

Midge larva

W. Patrick McCafferty, Aquatic Entomology, 1998: Jones and Bartlett Publishers; Sudbury, MA.
www.jbpub.com. Reprinted with permission

NEW MANAGEMENT
STRATEGIES AND
TECHNOLOGIES
OFFER HOPE
FOR THE FUTURE

Picking Up the Pieces

Summary

Students relate the challenges of restoring an altered watershed to piecing together a dismantled object.

Objectives

Students will:

- recognize how watersheds become altered.
- compare and contrast watersheds and objects before and after restoration.
- recognize the challenges of restoring altered watersheds.
- present findings about a watershed success story to the class in a symposium format.

Materials

- *Copy of the* **Cuyahoga Watershed Restoration Case Study** (or refer to www.cleanwater.gov/success for other watershed restoration case studies)
- *One object with a few parts to dismantle* (e.g., picture frame, pen)
- *Hand tools* (screwdrivers, wrenches, pliers, hammers)
- *6 assorted objects with multiple parts* (e.g., clocks, telephones, radios, etc.)
- *Optional:* Puzzles of varying difficulty can be substituted for the assorted objects listed above.

Background

Broadly defined a watershed is an area of land and the water that drains from it into a common outlet. More specifi-

Grade Level:
6–12

Subject Areas:
Environmental Science, Government, Social Studies

Duration:
Preparation: 30 minutes

Activity: Two 50 minute periods

Setting:
Classroom

Skills:
Organizing, Analyzing, Applying, Problem Solving

Vocabulary:
watershed, restoration

cally, watersheds are complex and diverse arrangements of physical and biological components. The water quality within a watershed is a reflection of the land uses and practices in that watershed. When watersheds are altered or impaired, the health of the water within the watershed is also impaired. Land uses that cause impairment or erosion of the land will also cause impairment of the water quality as the sediment typically runs off directly into an adjacent waterway.

While diversity and complexity are crucial to the strength and resiliency of natural systems, it is this same complexity that makes restoring such systems very difficult once they are altered or destroyed. When simpler systems break down or are damaged, they can be taken apart and put back together with relative ease. The simple systems found in mechanical objects such as radios, clocks, bicycles or puzzles are good examples. In a system as complex as an altered watershed, restoration becomes more difficult.

Restoration of any system becomes even more complicated when pieces of the system have been lost. For example, if you attempt to repair a broken clock when a key component is lost or damaged, the clock can be reassembled, but it may never work as it once did. Although the clock is put back together, its system remains altered. The same is true of a watershed where key biological or physical features (e.g., wetlands, habitat, aquatic insects, etc.) are lost. Restoration efforts can repair some damage but cannot restore such a complex system to its original state.

Much of the damage done to our watersheds happened at a time when there was little awareness of the ramifications of certain polluting endeavors. Activities such as mining, construction, logging, and dumping of untreated wastewater (primarily point source pollutants with identifiable sources) contaminated ground water, destroyed wildlife and increased both erosion and saltwater intrusion. At that time the main concern was not to preserve water quality but rather to meet the growing demand for natural resources; restoring altered land and water resources was given little consideration.

Only in the last few decades have those responsible for degrading these resources been required to mitigate damage or restore impacted sites. Despite this progress however, damaging impacts on watersheds continue, often from activities that each of us does every day. Most of our water pollution today comes from nonpoint source pollutants. These pollutants, which have no identifiable source, include runoff from oily streets and parking lots, fertilizer-rich city parks and lawns. These runoff pollutants can be lessened by use of simple practices such as maintaining or planting buffer strips along streams, preserving wetlands, and properly applying fertilizers and pesticides.

As times have changed, so have people's attitudes about preserving and restoring the environment. Laws requiring land reclamation and protection have been in effect for several decades. These shifts in societal attitudes, along with new technologies and greater financial resources, now make possible the restoration of previously forgotten or long-avoided sites. Many areas, sites once considered permanently altered or contaminated beyond hope, now have a chance at recovery. Government agencies, mining companies, logging firms, construction operations, developers, farmers, and landowners are now involved in countless restoration projects nationally and internationally. The farmer who plants vegetation to plug a draining wetland is taking part. The industry that cleans up the soil and water of a polluted waste pond and turns it into a park makes a difference. The government agency that begins the process of restoring a hazardous waste site with a plan to turn it into a wildlife refuge makes a contribution.

Restoration of contaminated or altered water environments is an example of both human ingenuity and Earth's ability to heal past wounds. However, restoration can rarely return a system to its original state and is no substitute for protecting and preserving unaltered natural systems. But when damage to natural systems occurs, restoration is a promising solution. It helps reclaim the past and offers young people hope and optimism for the future.

Cuyahoga River on fire June 22, 1969. Courtesy: Cuyahoga National Historical Park.

Helping restore a watershed is available to just about anyone. An individual, a family, a neighborhood, a student, a classroom, or a school system can help identify, guide and mobilize resources for a restoration project.

Procedure
Warm Up
Part I
Share with students the *Cuyahoga Watershed Restoration Case Study* or summarize the highlights with them. (You can also have students find a restoration story closer to home. A good place to start is at www.cleanwater.gov/success.)

a. Have students list the ways the watershed was altered.

b. List the ways the watershed is being restored (soil erosion abatement or flood control).

c. Ask students what indicators demonstrate the watershed is becoming healthier? (Steel head trout adults are now using Cuyahoga River.)

Part II
Show students an object with a few simple parts (e.g., picture frame, pen, mechanical pencil). Ask students to speculate how easy or difficult it might be to take the object apart and put it back together again. List any barriers to or difficulties with re-assembly that they can foresee. Dismantle the object and put it back together in front of them, explaining that the object represents a simple system that was altered and then restored to its former condition.

Next they will be challenged to restore more complex systems representing watersheds.

The Activity
Part I
1. **Divide students into six groups and assign each group a number. Provide each group with an item to dismantle (e.g., clock, radio, telephone, old microscope, etc.).** Explain that each object represents a watershed, and their task is to change it by dismantling it into small pieces.

2. **After dismantling is complete, place the pieces in a container marked with the group's number.**

3. **Without the other groups overhearing you, go around to each group and have them remove several parts from their dismantled piles to represent key pieces of the watershed that may have been lost (e.g., wetlands, habitat, aquatic insects, etc.).** These pieces should be placed in a container, labeled with their group number, and given to the instructor for use in a later discussion.

4. **Switch the containers and distribute them so that each group has a different object than the one they dismantled.**

5. **Their next task is to completely reassemble the new object.** Explain the reconstruction of their object will represent the restoration of an altered watershed. The actual tools (screwdriver or pliers) that the students need to

reconstruct their objects will represent the tools of watershed restoration. Ensure that there is *not* enough of each tool for all groups to reconstruct their object. This shortage represents the difficulties of finding the proper tools and experienced people needed for watershed restoration projects.

6. **After the students have reassembled their objects, return the missing pieces that were removed earlier.** How integral to the objects performance were those missing pieces?

Part II: Upper Level Option
1. **Working in small groups, have students complete the *Watershed Restoration Worksheet.*** To complete the worksheet, students should choose a watershed restoration case study from the Clean Water Action Plan (CWAP) Watershed Success Stories Web page (www.cleanwater.gov/success). There may be other watershed success stories available for local watersheds and you can substitute them for ones available on the CWAP Web page.

2. **Once they have identified their chosen watershed success story and completed their worksheets, have each group present their findings to the class.** Instruct students that they are to present their findings as if they were a team of scientists presenting their findings at a symposium.

3. **Create a rubric for their presentation and grade the group presentations accordingly.**

Wrap Up

Ask students to share their experiences of trying to restore their object back to its original state. Were they successful? Discuss the relationship of the activity to real-life restoration projects. Have students summarize how and why watersheds are changed and why they are difficult to restore. Ask how their restoration efforts were affected by the missing pieces.

While recognizing our continuous need for natural resources, ask students what can be done to help healthy watersheds maintain their integrity. What practices can they do in their daily lives that will help protect the watershed in which they live? What will make restoration of an altered watershed easier?(Photos of the watersheds taken before alteration, baseline data on water quality and plant and animal species, a water quality monitoring plan to track changes in the watershed, land-use practices that promote a healthy watershed—including erosion controls, buffer strips, etc.)

Assessment

Have students:
- recognize that watersheds can become altered (*Warm Up*).
- compare and contrast watersheds and objects before and after restoration (*The Activity*).
- recognize the challenges of restoring an altered watershed (**Steps 1-6**).
- present findings about a watershed success story to the class in a symposium format (*Upper Level Option*).

Extensions

Have student groups conduct the activity using only one hand each. Let this handicap represent the difficulty of restoring a watershed with limited tools and the importance of cooperation in such efforts.

Have students research some of the tools used in watershed restoration. What are the actual costs for these watershed restoration tools? Have students research various watershed restoration projects and list the tools and their costs.

Have students research other water-related restoration projects underway locally, regionally, or nationally. There are several government agency Web sites that have links to restoration success stories (see **Resources**).

Students could identify and undertake a water-related restoration project in their area. To determine the feasibility of the project they will need to consider how they will: (1) establish a restoration goal; (2) formulate a restoration plan; (3) predict the difficulties they will encounter; (4) analyze costs; (5) determine a time frame and project results (e.g., illustrate the potential appearance of restored site); (6) provide for maintenance of the restored site. If the project proves feasible and students undertake restoration of a site, have them maintain a project diary or water log and circulate copies to other teachers and students. Students can give a feature story interview to the local newspaper. Help students write a story about the project and share it with their local newspaper and government agencies.

Resources

Clean Water Action Plan. 2000. *CWAP Watershed Success Stories.* Retrieved February 17, 2002, from the Web site: http://www.cleanwater.gov/success

Galatowitsch, S.M., and A. van der Valk. 1994. *Restoring Prairie Wetlands: An Ecological Approach.* Ames, Iowa: Iowa State University Press.

New York Governor's Office. 1998. Governor Announces $25.6 Million for Hudson River Counties. Retrieved February 17, 2002, from the Web site: http://www.state.ny.us / governor/press/oct08_2_98.htm.

Notes:

Watershed Restoration Worksheet

1. Describe the watershed—its location, length, drainage area, and prominent features.

2. Identify the pollution history of the watershed. Why does the watershed need restoring?

3. List the mission of the watershed RAP—Remedial Action Plan—(plan for restoring the watershed).

4. Who is doing the restoration work in this watershed?

5. List the projects used to restore the watershed.

6. What community activities are aiding restoration efforts in the watershed?

7. List indicators or examples of improved environmental quality in the watershed.

Going Underground

Summary

Students create ground water models to observe the interaction between ground water and surface water and to study how this relationship affects ground water pollution.

Objectives

Students will:

- determine the relationship between ground water and surface water.
- create a model to represent ground water and surface water.
- identify the water table, saturated zone, unsaturated zone, ground water and surface water on their model.

Materials

Warm Up

- *A large bowl or plastic container*
- *Paper cups* (two)
- *Sand* (one cup)
- *Powdered drink mix*
- *Water*

Activity

- *Plastic 2-liter bottle* (one per group)
- *Paper cups* (two per group)
- *Clay* (enough to thinly cover the bottoms of the bottles)
- *Gravel* (about two cups per group)
- *Sand* (about two cups per group)
- *Powdered drink mix* (red or purple)
- *Pump from lotion or soap dispenser* (one per group)
- *Spoon or other device to dig up*

Grade Level:
6–12

Subject Areas:
Earth Science, Geology

Duration:
Preparation: 30 minutes

Activity: 50 minutes

Setting:
Classroom

Skills:
Observe, Synthesize, Create, Model, Label, Quantify

Vocabulary:
remediate, ground water, surface water, water table, saturated zone, unsaturated zone, aquifer

sand (one per group)
- *Copies of **Ground Water Scenarios** Student Copy Pages* (one copy, cut into four)
- *Copies of **Ground Water Worksheet** Student Copy Pages* (one per group)
- *Food coloring* (two drops per group)
- *Cheese cloth* (two 2-inch squares per group)
- *Rubber bands* (1 per group)
- *Large cup, bottle, or other device to collect water* (one per group)

Background

Where does water go when it rains? When rainwater falls to the Earth, some evaporates; some flows over land as runoff to eventually join rivers, lakes or oceans; some falls directly onto rivers, lakes or oceans; and some soaks into the earth to become ground water.

As rainwater soaks into the ground, it slowly percolates by the force of gravity down through various types and layers of soil and rock. These soils have varying degrees of permeability (the ability to let water pass freely through them). Permeable soils are those that allow water to pass through with relative ease. Impermeable soils are more resistant. Sand and gravel are very permeable, for instance, because they contain relatively large spaces between particles of soil called pore spaces. Impermeable soils, on the other hand, do not contain pore spaces and, therefore, effectively block water from flowing through them. Clay and granite are two good examples of impermeable materials.

Water percolates downward until it reaches an impermeable layer, then begins to accumulate and fill the pore spaces in the permeable layer above. When enough water accumulates and pore spaces are completely filled with water, the area is considered saturated. Above this saturated zone, pore spaces may contain some water but also some air. This layer is known as the unsaturated zone. The surface where the saturated zone and the unsaturated zone meet is called the water table. This phenomenon, which stores and transmits ground water, is called an aquifer.

Aquifers are extremely important sources of fresh water. About half of the United States' fresh water supply comes from ground water stored in aquifers (Environmental Concern, Inc. and The Watercourse, 1995), and fresh water wells draw water directly from the nearest aquifer.

Since the water in aquifers was once on the surface of the Earth, it is subject to pollution just as surface water bodies are. Water percolating through the soil can carry pollutants along with it. Percolation of rainwater through soil to create ground water isn't the only instance of surface water interacting with ground water. Another example is seen when the water table intersects the Earth's surface to form lakes. Such ready interaction between surface and ground waters puts fresh water sources at risk, making it critical to protect both from pollution.

A boy lifting a pump off of a well so the ground water can be monitored. Courtesy: Dr Stephan Custer, Montana State University

Procedure

Warm Up

Ask students to brainstorm all the places that rain goes after it falls to the Earth. Some rain may evaporate, some may fall on oceans or lakes, some may fall directly on streams or rivers, some may run over the land (runoff) to join streams, rivers, lakes, and oceans, and some may soak into the ground to become ground water.

There is an important interaction between surface water (lakes, oceans, streams, rivers, ponds) and ground water. Some rainwater percolates through the soil to form ground water. Likewise, anything applied to the ground can percolate through the soil to join ground water, including pollutants. Read this scenario to students:

Rachel just finished changing the oil in her car. She decided to dispose of the old oil in the back yard, away from her garden, the doghouse, and where her children play. Does this oil pose a threat if it is far away from any human or animal activities?

To illustrate the interaction between surface water and ground water, fill a bowl with gravel until it is a few inches from the top. Fill to the same level with water. Next poke *small* holes in the bottom of two paper cups. Place about an inch of sand in one of the cups.

Explain to students that the sand in the paper cup represents soil on the surface of the Earth. Set the cup in the bowl, resting on the gravel and water. The bowl represents an aquifer where ground water is stored. Place about a teaspoon of red or purple powdered drink mix on top of the sand in the cup. This mix represents the oil Rachel poured out in her backyard.

Tell students that it started to rain shortly after Rachel poured out the oil. While holding the empty paper cup (with holes in the bottom) a few inches above the paper cup with sand and "oil", pour water into the top cup. This simulates rain falling onto the sand and "oil". The rainwater should then percolate through the sand and "oil", into the "ground water" (bowl with gravel and water).

As shown in the demonstration, how does water falling on the surface of the Earth impact ground water?

Does the oil remain in the sand or does it travel into the ground water? Even though Rachel made an effort to dispose of her oil away from any human and animal activities, does it still affect humans and animals? Ask the students if humans or animals use ground water. They have learned that surface water can influence ground water, but can ground water influence surface water?

The Activity

1.   **Students simulate the interaction between ground water and surface water in mini ground water models.** Begin by distributing the following items to each group (ideally four or eight groups total):

- *plastic 2-liter bottle*
- *a pump from a lotion or soap dispenser*

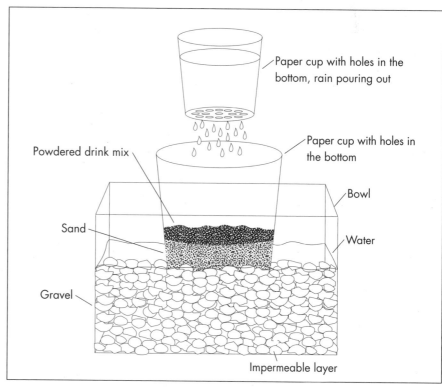

Powdered drink mix
Paper cup with holes in the bottom, rain pouring out
Paper cup with holes in the bottom
Sand
Bowl
Water
Gravel
Impermeable layer

- *clay* (enough to thinly cover the bottom of the bottles)
- *gravel* (about two cups)
- *sand* (about two cups)
- *a spoon or other device to move sand*
- *water*
- *two drops of food coloring*
- *two paper cups*
- *cheese cloth*
- *rubber band*
- *a large cup, bottle or other device to collect water*
- *a device to measure volume* (measuring cup, beaker, etc.)

2. **Have students cut several inches off the top of the plastic 2-liter bottle so the bottle is slightly shorter than the length of the pump.** The pump will stand in the bottom of the bottle to pump water out.

3. **As a class, assemble the ground water models. First, place a thin layer of clay on the bottom of the bottle to represent the impermeable layer. Then, pour two cups of gravel into the 2-liter bottle. Next, pour about two cups of sand on top of the gravel.** If large gravel is used, you may need to add more sand as it may take more to fill in the larger pore spaces between the gravel.

4. **To demonstrate how ground water accumulates, have students pour water into the paper cup (with holes in the bottom) while holding it over the model of the ground (cut 2-liter bottle) so that it percolates through the soil.** This represents rainwater falling on the ground, and percolating through the soil to form ground water. Continue to fill the bottle until the water is just past the top of the gravel, but not to the top of the sand. If students look at the side of their model, they should be able to see and monitor the water. What is the term for the top of the water level? (Water table.) Above the water table is the unsaturated zone and below the water table is the saturated zone where water fills all the pore spaces between grains of gravel and sand.

5. **Instruct the groups to make a lake in the ground water model by scooping the sand from one side of the model and piling it on the other side of the model. Dig down until you reach gravel.** A small lake should form above the gravel. If more water is needed to make the lake, add some at this time.

6. **Discuss how the water table can sometimes intersect the Earth's surface.** When this happens, ground water and surface water freely interact. The shore of the lake represents this in your model. Beneath the pile of sand, the water table should be visible through the plastic bottle—this is ground water. The water seen from the surface on the other side of the model is surface water.

7. **Cover the bottom of the lotion pump with cheesecloth; securing it with a rubber band.** Insert the lotion pump into the mound of sand in the model. Ask students what this pump could represent (a well). How deep do they need to insert their well to ensure that it pumps ground water from their aquifer?

8. **Distribute a different scenario from the *Ground Water Scenarios* to each group.** Ask them to follow the instructions in their scenario and then complete the *Ground Water Worksheet.*

9. **After all groups have completed the *Ground Water Worksheet,* have each group present their scenario and results to the class.** Have them address where they placed the pollution. Did it contaminate surface water, ground water, or both?

Wrap Up

Where does rainwater go once it has landed on Earth? (Runoff, oceans/lakes/rivers, evaporate, or ground water.) How does it accumulate in the ground? What causes it to begin accumulating? (Impermeable layer, i.e., clay or rock.) What part of their models represents the permeable and impermeable layers? (Sand/gravel, plastic bottom of the bottle, respectively.) What part of their model represents an aquifer? (Saturated zone.)

If you pollute surface water, do you also pollute ground water? Vice versa? You used a pump to remove the polluted water from your model. Can you think of other times water is pumped from the ground? Ground water is a component of many municipal drinking water supplies. Rural homes often have wells that supply ground water for their household water uses. Is your home supplied by well or city water? What other sources of water might be used in urban areas? What can cities do to prevent pollution in their water supply?

Assessment

Have students:
- determine the relationship between ground and surface water (*Warm Up*).
- create a model to represent ground and surface water (**Steps 2-3**).
- identify the water table, saturated zone, unsaturated zone, ground water and surface water from their model (**Step 4-6**).

Extensions

Determine how much water it takes to remove all of the pollution. Use the pump to remove the polluted water from your model. Be sure to place the bottom of the pump in the gravel. Pump the water into a large cup or bottle. Once the water in your ground water model is clear,

you have removed all of the pollution. Your model is not clean if you have simply removed all of the water! You must continue to add water until the water remains clear. Record the volume of water needed to clean up two drops of "pollution" from your ground water model.

Explain that according to the Virginia Department of Conservation and Recreation, one gallon of used motor oil can pollute up to 2 million gallons of water. This explains why it takes so much water to clean just two drops of "pollution."

Have students determine the source(s) of their municipal water supply.

Contact a local geologist or hydrologist and ask them to help the students create ground water models that represent the types of soil layers found in your area. Does local geology influence the interaction between ground and surface water? Does this affect the time it

takes to remediate the pollution in the model?

Testing Kit Extensions

Pollute the water in the model using nitrate-rich fish food or fertilizer. Use the Healthy Water, Healthy People testing kit to measure the amount of nitrates before and after polluting. Continue flushing water through the ground water model until the level of nitrates has returned to its original pre-polluted level. Confirm this by retesting the water for nitrates using the Healthy Water, Healthy People testing kit.

Resources

Environmental Concern, Inc. and The Watercourse. 1995. *WOW! The Wonders of Wetlands.* Environmental Concern, Inc. and The Watercourse.

Simons, B., and T. Wellnitz. 2000. *Science Explorer: Earth's Waters.* Upper Saddle River, NJ.: Prentice Hall.

Notes:

Ground Water Scenarios

Scenario 1: An old underground gas tank on your property is leaking waste into the ground water.

1. Dig a small hole about two inches deep in the sandy side of your ground water model.
2. Place two drops of food coloring in the hole to represent the leaking waste.
3. Re-cover the hole with sand.
4. Later it rains. To simulate this, put water in the paper cup with holes in the bottom. Repeat four times.
5. Answer questions from the Ground Water Worksheet.

Scenario 2: Your neighbor continues to dispose of his used motor oil in his back yard.

1. Place two drops of food coloring on the surface of the sandy side of your model to simulate the oil disposal.
2. Later it rains. To simulate this, put water in the paper cup with holes in the bottom. Repeat four times.
3. Answer questions from the Ground Water Worksheet.

Scenario 3: It has just been revealed that a factory has been discharging water contaminated with hazardous waste into the bay on which you live. The factory is being reprimanded, but it will take time to remove the hazardous waste.

1. Place two drops of food coloring in the lake of your model to simulate hazardous waste in the bay.
2. Later it rains. To simulate this, put water in the paper cup with holes in the bottom. Repeat four times.
3. Answer questions from the Ground Water Worksheet.

Scenario 4: You live in a well-established neighborhood on a lakeshore. You and your neighbor have well-manicured lawns extending almost to the beach. You both fertilize your lawns regularly to maintain their appearance.

1. Place two drops of food coloring on the lakeshore, two inches apart. This represents you and your neighbor fertilizing your lawns.
2. Later it rains. To simulate this, put water in the paper cup with holes in the bottom. Repeat four times.
3. Answer questions from the Ground Water Worksheet.

Ground Water Worksheet

1. Label the following items on the ground water model picture:
   a. water table
   b. saturated zone
   c. unsaturated zone
   d. ground water
   e. surface water

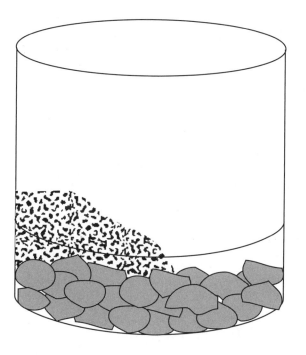

2. After adding the pollution, list the places you can observe it in your model.

3. Pump a little water out of the ground water using your lotion pump, or simulated well. What color was the water? Depending on the type of pollution represented in your scenario, is the water from your well safe to drink now that it contains this pollutant? (NOTE: Do not drink the water!)

4. In a paragraph or illustration, summarize how precipitation can become ground water, and how pollution on the land's surface can also pollute ground water.

Appendixes

Activities by Subject

Activity	Computer Science	Health	History	Language Arts	Math/Metrics	Biology/Life	Chemistry	Earth	Ecology	Environmental	General	Geography	Geology	Microbiology	Physics	Ethics	General	Government	Political Science	Technology
						Science										**Social Studies**				
Benthic Bugs & Bioassessment p.154					■	■				■	■									
Carts and Horses p. 42	■	■	■	■	■	■	■	■	■	■	■	■	■	■	■	■	■	■	■	■
Footprints on the Sand p. 90			■				■	■	■	■		■	■		■			■		
From H to OH! p. 15			■				■	■			■	■								
Going Underground p. 187				■				■			■		■							■
Grab a Gram p. 29		■		■		■	■	■		■								■		■
Hitting the Mark p. 49				■			■			■	■									
Invertebrates as Indicators p. 174				■		■			■	■										
It's Clear to Me! p. 10								■				■				■				■
Life and Death Situation p. 125		■				■				■	■	■		■						
Looks Aren't Everything p. 99				■	■						■	■		■						
Mapping It Out: Water Quality Concept Mapping p. 6	■	■	■	■	■	■	■	■	■	■	■	■	■	■	■	■	■	■	■	■
Multiple Perspectives p. 55	■	■		■	■					■	■					■	■	■		■
Picking up the Pieces p. 182				■		■		■	■	■	■						■	■		■
Pollution–Take It or Leave It! p. 21				■				■		■	■					■				
Setting the Standards p. 107			■	■			■	■		■	■			■				■	■	
A Snapshot in Time p. 61				■	■			■			■	■								
Stone Soup p. 35						■		■		■	■									
A Tangled Web: Conducting Internet Research p. 1	■			■		■	■			■										■
There is No Point to This Pollution! p. 136					■			■	■	■	■	■		■		■	■	■		
Turbidity or not Turbidity: That is the Question! p. 83					■		■			■	■				■					■
Washing Water p. 145		■				■	■			■	■							■		
Wash It Away p.121				■	■			■			■								■	■
Water Quality Monitoring: From Design to Data p. 70	■			■			■	■		■	■									
Water Quality Windows p. 164						■	■		■		■	■								

Activities by Skill

Activity	Apply	Analyze	Calculate	Cause and Effect	Classify	Compare and Contrast	Combine	Communicate	Create	Debate	Demonstrate	Distinguish	Evaluate	Experiment Design	Gather/Collect Data
Benthic Bugs and Bioassessment p. 154								■							■
Carts and Horses p. 42	■	■						■	■					■	■
Footprints in the Sand p. 90	■														
From H to OH! p. 15								■							
Going Underground p. 187									■						
Grab a Gram p. 29	■	■	■			■									
Hitting the Mark p.49	■	■												■	■
Invertebrates as Indicators p. 174		■			■										
It's Clear to Me! p.10						■	■					■			
Life and Death Situation p. 125		■											■		■
Looks Aren't Everything p. 99		■	■										■		■
Mapping it Out: Water Quality Concept Mapping p. 6	■	■											■		■
Multiple Perspectives p. 55		■								■			■		■
Picking up the Pieces p. 182	■	■													
Pollution–Take it or Leave It! p. 21		■			■										■
Setting the Standards p. 107		■													
A Snapshot in Time p. 61		■						■							■
Stone Soup p. 35						■					■		■		
A Tangled Web: Conducting Internet Research p. 1		■											■		■
There is No Point to This Pollution! p. 136		■				■				■					
Turbidity or Not Turbidity: That is the Question! p. 83						■									
Washing Water p.145															■
Wash It Away p. 121	■			■		■		■			■	■	■		■
Water Quality Monitoring: From Design to Data p. 70		■											■		■
Water Quality Windows p. 164					■										

Infer	Interpret	Investigate	Label	Manipulate	Match	Measure	Model	Organize/Mapping/Graph	Observe	Plan	Predict	Present	Problem Solve	Procedure Design	Quantify	Rank	Research	Record	Simulate	Synthesize	Whole-Body Movement	Quantify
	■							■														
		■							■		■						■					
	■						■										■					
	■							■								■						
			■				■													■		■
	■	■				■		■														
	■											■		■								
	■										■											
								■										■				
■								■														
	■											■										
	■							■				■										
■	■	■										■					■					
								■					■									
	■							■														
	■																		■	■		
	■							■				■										
		■					■															
	■						■															
	■							■														■
	■	■				■						■									■	
				■		■	■			■												
■	■	■		■			■		■						■	■		■	■			■
	■											■										
■	■				■								■									

Topics / Activity	Educational Principals		What is Water Quality?				Illustrating Scientific Processes			
	Concept Mapping	Internet Searches & Evaluation	Measurement	Hydrologic Cycle	Solutions & Mixtures	Understanding Specific Parameters	Inquiry/Scientific Method/Study Design	Accuracy & Precision	General Human Health	Differing Views of Data
Benthic Bugs and Bioassessment p. 154							■			
Carts and Horses p. 42									■	
Footprints on the Sand p. 90							■		■	
From H to OH! p. 15					■			■	■	
Going Underground p. 187				■			■		■	
Grab a Gram p. 29			■		■		■	■	■	
Hitting the Mark p. 49							■	■		
Invertebrates as Indicators p. 174						■	■			
It's Clear to Me! p. 10					■		■			
Life and Death Situation p. 125						■	■		■	
Looks Aren't Everything p.99						■	■		■	
Mapping It Out: Water Quality Concept Mapping p. 6	■						■			
Multiple Perspectives p. 55							■		■	■
Picking up the Pieces p. 182							■			
Pollution–Take It or Leave It! p. 21				■			■			
Setting the Standards p. 107			■			■	■		■	■
A Snapshot in Time p. 61						■	■			■
Stone Soup p. 35					■	■	■			
A Tangled Web:Conducting Internet Research p. 1		■					■			
There is No Point to This Pollution! p. 136						■	■			
Turbidity or not Turbidity– That is the Question! p. 83			■	■	■	■	■			
Washing Water p. 145			■				■		■	
Wash it Away p. 121						■	■		■	
Water Quality Monitoring: From Design to Data p. 70						■	■			■
Water Quality Windows p.164							■			

Data Analysis	Water Quality is Monitored & Interpreted		Healthy Communities & Individuals				Water is Managed to Prevent Degradation				Ecological Effects of Water Quality		Hope for the Future
Data Analysis	Water Quality Monitoring	Aquatic Invertebrate Sampling	Drinking Water	Contamination Heavy Metal/Microbial	Toxicity	Water Quality Standards	Nonpoint Source Pollution	Water & Wastewater Treatment	Ground Water	Plants & Animals	Aquatic Invertebrates	Restoration	Action projects
■	■	■					■	■		■	■		■
		■											
	■			■			■	■		■			■
■		■			■			■					
■	■		■	■				■	■				
■			■		■	■	■	■					
■	■						■	■					
■	■	■					■				■	■	■
■			■										
■			■	■			■			■			
■			■	■			■	■					■
			■	■			■			■		■	
■	■		■				■	■	■		■		
■	■						■	■		■		■	■
								■	■	■		■	
■	■		■	■		■	■	■					
■	■						■	■					■
							■						
			■				■	■					■
■	■			■		■	■			■			■
■	■		■				■	■		■		■	
■													
			■	■		■	■	■					
■	■						■	■					■
■					■					■			

Content Standards, Grades 5-8

■ Completely meets Standards
□ Partially meets Standards

Page	Activity	Systems, order and organization	Evidence, models, and explanation	Change, constancy, and measurement	Evolution and equilibrium	Form and function	Abilities necessary to do scientific inquiry	Understandings about scientific inquiry	Properties and changes of properties in matter	Motions and forces	Transfer of energy
		Unifying Concepts and Processes					Science as Inquiry		Physical Science		
154	Benthic Bugs and Bioassessment	■	□	■		□	■	■			
42	Carts and Horses	□	■	□		□	■	■			
90	Footprints on the Sand		□						□		
15	From H to OH!	□	■	□			□	□	■		
187	Going Underground		■	□		□	□	□			
29	Grab a Gram	■		□			□				
49	Hitting the Mark		■	□		□	■	□			
174	Invertebrates as Indicators	■	■	■	□	■	□	■			
10	It's Clear to Me!	■	■	□			□	□	□		
125	Life and Death Situation		□				□	□			
99	Looks Aren't Everything		□				□	□			
6	Mapping It Out: Water Quality Concept Mapping	■				□	■	■			
55	Multiple Perspectives					■		□			
182	Picking Up the Pieces	□		□		□		□			
21	Pollution–Take It or Leave It!	□	□		□				□		
107	Setting the Standards	□	■	□			■	■			
61	A Snapshot in Time	□	□	■			□	□			
35	Stone Soup		□						□		
1	A Tangled Web: Conducting Internet Research	□					□	□			
136	There is No Point to This Pollution!		■	□			□	□			
83	Turbidity or Not Turbidity: That is the Question!		□	□							
145	Washing Water	□		□			□	□			
121	Wash it Away		□				□	□			
70	Water Quality Monitoring; From Design to Data	□	□	■		□	■	■			
164	Water Quality Windows					□	□				

Life Science					Earth and Space Science			Science and Technology		Science in Personal and Social Perspectives					History and Nature of Science		
Structure and function in living systems	Reproduction and heredity	Regulation and behavior	Populations and ecosystems	Diversity and adaptations of organisms	Structure of the earth system	Earth's history	Earth in the solar system	Abilities of technological design	Understandings about science and technology	Personal health	Populations, resources, and environments	Natural hazards	Risks and benefits	Science and technology in society	Science as a human endeavor	Nature of science	History of science
□	□		■	□							□				□	□	
								■	□						□	■	□
										□	■	□	■	□	□	□	
									□							■	
					□					□	■	□	□	□		□	
									□						□	□	
															□	□	
■	□	■	■	■							□						
															□	□	
			□							■	■		■	□	□	□	
			□							■	■		□	□	□	□	
															■	□	
										□	■		□	■	■	□	
			□	□				□	□					□	□	□	
			□								□						
					□			□	□	■	□		□	■	□	□	
											□				□	□	
								□	□								
										□	□		□	□	□	□	
											□						
								□	□	□	□			□			
			□							■	■		■	□	□	□	
											□			□	□	□	
□			■	□							□						

Content Standards, Grades 9-12

■ Completely meets Standards
□ Partially meets Standards

Page	Activity	Unifying Concepts and Processes					Science as Inquiry		Physical Science				
		Systems, order and organization	Evidence, models, and explanation	Change, constancy, and measurement	Evolution and equilibrium	Form and function	Abilities necessary to do scientific inquiry	Understandings about scientific inquiry	Structure of atoms	Structure and properties of matter	Chemical reactions	Motions and forces	Conservation of energy and increase in disorder
154	Benthic Bugs and Bioassessment	■	□	■		□	■	■					
42	Carts and Horses	□	■	□			■	■					
90	Footprints on the Sand		□							□			
15	From H to OH!	□	■	□			□	□		■	□		
187	Going Underground		■	□		□	□	□					
29	Grab a Gram	■		□						□			
49	Hitting the Mark		■	□		□	■	□					
174	Invertebrates as Indicators	■	■	■	□	■	□	■					
10	It's Clear to Me!	■	■	□			□	□			□	□	
125	Life and Death Situation		□				□	□			□	□	
99	Looks Aren't Everything		□				□	□					
6	Mapping It Out: Water Quality Concept Mapping	■				□	■	■					
55	Multiple Perspectives					■		□					
182	Picking Up the Pieces	□		□		□		□					
21	Pollution–Take It or Leave It!	□	□		□					□			
107	Setting the Standards	□	■	□			■	■					
61	A Snapshot in Time	□	□	■			□	□					
35	Stone Soup		□									□	
1	A Tangled Web: Conducting Internet Research	□					□	□					
136	There is No Point to This Pollution!		■	□			□	□					
83	Turbidity or Not Turbidity: That is the Question!		□	□									
145	Washing Water	□		□			□	□					
121	Wash it Away		□				□	□					
70	Water Quality Monitoring: From Design to Data	□	□	■		□	■	■					
164	Water Quality Windows					□	□						

Interactions of energy and matter	The cell	Molecular basis of heredity	Biological evolution	Interdependence of organisms	Matter, energy, and organization in living systems	Behavior of organisms	Energy in the earth system	Geochemical cycles	Origin and evolution of the universe	Abilities of technological design	Understandings about science and technology	Personal and community health	Population growth	Natural resources	Environmental quality	Natural and human-induced hazards	Science and technology in local, national, and global challenges	Science as a human endeavor	Nature of scientific knowledge	Historical perspectives
			□	■	□	□						□	□	□	■	□	□	□	□	
										■	□							□	■	□
				□								□	□	■	■	□	■	□	□	
											□								■	
								■				□	□	■	■	□	□			
											□				□				□	
										□	□							□	□	
	■			■	■	■														
														□				□	□	
				□								■	□	■	■	□	■	□	□	
				□								■	□	■	■	□	□	□	□	
																		■	□	
												■	□	■	□		■	■	□	
				□						□	□			□			□	□	□	
					□			■						□	□					
								□		□	□	■		□	■	□	□	□	□	
														■	□			□	□	
																		□		
							□	□												
												□	□	□	□		□	□	□	
														□	□					
							□	□		□		□		□	□		□			
												■	□	□	□	□		□	□	
														□	■		□	□	□	
		□	□	□	□							□								

Cross Reference Between Project WET & Watercourse Publications

Activity Topics	Healthy Water, Healthy People	Project WET Curriculum and Activity Guide
Aquatic Invertebrate Sampling	Benthic Bugs and Bioassessment Pg. 154	Macroinvertebrate Mayhem Pg. 322
Scientific Method	Carts and Horses Pg. 42	
Nonpoint Source Pollution	Footprints on the Sand Pg. 90	A-maze-ing Water Pg. 219 The Pucker Effect Pg. 388 Sum of the Parts Pg. 267 Whose Problem Is It? Pg. 429 Wish Book Pg. 460
pH	From H to OH! Pg. 15	Where Are the Frogs? Pg.279
Drinking Water Contamination	Wash it Away Pg. 121	A Grave Mistake Pg.311 Poison Pump Pg.93
Ground Water	Going Underground Pg. 187	Get the Ground Water Picture Pg. 136 A Grave Mistake Pg. 311 The Pucker Effect Pg. 388
Understanding Measurement/Parts-Per-Million	Grab a Gram Pg. 29	Reaching Your Limits Pg. 344
Accuracy and Precision	Hitting the Mark Pg. 49	
Aquatic Invertebrates as Water Quality Indicators	Invertebrates as Indicators Pg. 174	Macroinvertebrate Mayhem Pg. 322
Solutions & Mixtures	It's Clear to Me! Pg. 10	What's the Solution? Pg. 54
Microbial Contamination	Life and Death Situation Pg. 125	Poison Pump Pg. 93 No Bellyachers Pg. 85 Super Sleuths Pg. 107
Microbial Contamination	Looks Aren't Everything Pg. 99	A-maze-ing Water Pg. 219 No Bellyachers Pg. 85 Poison Pump Pg. 93
Concept Mapping	Mapping It Out: Water Quality Concept Mapping Pg. 6	Check It Out! Pg. 3 Idea pools Pg. 7
Differing Views Of An Issue	Multiple Perspectives Pg. 55	Dilemma Derby Pg. 377 Hot Water Pg. 388 Perspectives Pg. 397 Water Court Pg.413
Watershed Restoration	Picking up the Pieces Pg. 182	Humpty Dumpty Pg. 316
Hydrologic Cycle	Pollution—Take It or Leave It! Pg. 21	Imagine! Pg. 157 The Incredible Journey Pg. 161 Thirsty Plants Pg. 116 Water Models Pg. 201
Water Quality Standards	Setting the Standards Pg. 107	Reaching Your Limits Pg. 344
Watershed Monitoring	A Snapshot in Time Pg. 61	Branching Out! Pg.129 Color Me a Watershed Pg. 223 Just Passing Through Pg. 166 Rainy-Day Hike Pg. 186
Alkalinity	Stone Soup Pg. 35	Where Are the Frogs? Pg. 279
Internet WebQuest & Web site Evaluation	A Tangled Web: Conducting Internet Research Pg. 1	
Nonpoint Source Pollution/Watersheds	There is No Point to This Pollution! Pg. 136	A-maze-ing Water Pg. 219 Sum of the Parts Pg. 267
Turbidity & Sediment	Turbidity or not Turbidity: That is the Question! Pg. 83	Capture, Store, and Release Pg.133
Water Treatment & Wastewater Treatment	Washing Water Pg. 145	Sparkling Water Pg. 348 Super Bowl Surge Pg. 353
Water Quality Monitoring & Study Design	Water Quality Monitoring: From Design to Data Pg. 70	Branching Out! Pg.129 Capture, Store, and Release Pg.133 Color Me a Watershed Pg. 223
Water Quality Effects On Plants & Animals	Water Quality Windows Pg. 164	Salt Marsh Players Pg.99

Cross Reference Between Project WET & Watercourse Publications/Cont.

Conserve Water Educators Guide	WOW! Wonders of Wetlands	Watershed Manager Educators Guide
	Water We Have Here Pg.174	
	Recipe for Trouble Pg.199	
		Sum of the Parts Pg. 114
Water Trouble on the High Plains (CS) Pg.260		
	Water Under Foot Pg. 204 How Thirsty Is the Ground Pg. 239	
	Wet 'n' Wild Pg. 99	
The Ins and Outs of Water Conservation Pg. 23	Hear Ye! Hear Ye! Pg. 253 Wetland Tradeoffs Pg. 285	
Conservation Choices Pg.115 Planning With Vision (CS) Pg. 274	Get Involved! Pg. 310	Choices and Preferences Pg. 72 Pass the Jug Pg. 100
The Blue Traveler Pg. 43	Water We Have Here? Pg.174	The Blue Traveler Pg. 15
		Seeing Watersheds Pg. 4 Snow and Tell Pg. 79 Streams of Data Pg. 57 Rainy-Day Hike Pg. 52
		Watershed Web Pg. 9
		River Reflections Pg. 107 Seeing Watersheds Pg. 4 Blue Beads Pg. 45
The Problem With Silt (CS) Pg.282	Treatment Plants Pg.120	
Adrift (CS) Pg.249 Operation Water Sense (CS) Pg.264	Recipe for Trouble Pg.199	
Your Hydrologic Deck Pg.161	Nutrients: Part II The Nutrient Trap Pg.188 Runoff Race Pg.210 Wetland in a Pan Pg.212	Streams of Data Pg. 57 Hydrologic Bank Account Pg. 95 Watershed Manager Pg. 120 Your Hydrologic Deck Pg.124 Color Me a Watershed Pg. 157
Alligators, Epiphytes, and Water Managers Pg.75	Water Purifiers Pg. 215	

Activities by Grade Level

Activity	6-8	9-12
Benthic Bugs and Bioassessment p. 154	■	■
Carts and Horses p. 42	■	■
Footprints on the Sand p. 90	■	■
From H to OH! p. 15	■	
Going Underground p. 187	■	■
Grab a Gram p. 29	■	■
Hitting the Mark p. 49	■	■
Invertebrates as Indicators p. 174	■	
It's Clear to Me! p. 10	■	■
Life and Death Situation p. 125	■	■
Looks Aren't Everything p. 99	■	
Mapping It Out; Water Quality Concept Mapping p. 6	■	■
Multiple Perspectives p. 55	■	■
Picking up the Pieces p. 182	■	■
Pollution–Take It or Leave It! p. 21	■	
Setting the Standards p. 107		■
A Snapshot in Time p. 61	■	■
Stone Soup p. 35	■	■
A Tangled Web: Conducting Internet Research p. 1	■	■
There is No Point to This Pollution! p. 136	■	■
Turbidity or not Turbidity: That is the Question! p. 83	■	
Washing Water p. 145	■	■
Wash It Away p. 121	■	
Water Quality Monitoring: From Design to Data p. 70		■
Water Quality Windows p. 164		■

Metric and Standard Measurements
Metric Conversions

When You Know	Multiply By	To Find	When You Know	Multiply By	To Find
Length			**Volume**		
Inches (in.)	2.5	Centimeters (cm)	Teaspoons (tsp.)	5.0	Milliliters (ml)
Feet (ft.)	30.0	Centimeters (cm)	Tablespoons (tbs.)	15.0	Milliliters (ml)
Yards (yd.)	0.9	Meters (m)	Fluid ounces (fl.oz.)	30.0	Milliliters (ml)
Miles (mi.)	1.6	Kilometers (km)	Cups (c)	0.24	Liters (l)
Centimeters (cm)	0.4	Inches (in.)	Pints (pt.)	0.47	Liters (l)
Meters (m)	3.3	Feet (ft.)	Quarts (qts.)	0.95	Liters (l)
Meters (m)	1.09	Yard (yd.)	Gallons (gal.)	3.8	Liters (l)
Kilometers (km)	0.6	Mile (mi.)	Cubic feet (ft.^{3})	0.03	Cubic meters (m^3)
			Cubic yards (yd.^{3})	0.76	Cubic meters (m^3)
Area			Milliliters (ml)	0.2	Teaspoons (tsp.)
Square inches (in.^{2})	6.5	Square centimeters (cm^2)	Milliliters (ml)	0.7	Tablespoons (tbs.)
Square feet (ft.^{2})	0.09	Square meters (m^2)	Milliliters (ml)	0.3	Fluid ounces (fl.oz.)
Square yards (yd.^{2})	0.84	Square meters (m^2)	Liters (l)	4.2	Cups (c)
Square miles (mi.^{2})	2.6	Square kilometers (km^2)	Liters (l)	2.1	Pints (pt.)
Acre (a.)	0.4	Hectares (ha)	Liters (l)	1.06	Quarts (qts.)
Square centimeter (cm^2)	0.16	Square inches (in.^{2})	Liters (l)	0.26	Gallons (gal.)
Square meter (m^2)	10.8	Square feet (ft.^{2})	Cubic meters (m^3)	35.0	Cubic feet (ft.^{3})
Square meter (m^2)	1.2	Square yards (yd.^{2})	Cubic meters (m^3)	1.3	Cubic yards (yd.^{3})
Square kilometer (km^2)	0.4	Square miles (mi.^{2})			
Hectare (ha)	2.5	Acres (a.)	**Temperature**		
			Degrees Celsius ($^{\circ}$C)	(9/5 x $^{\circ}$C) +32	Degrees Fahrenheit ($^{\circ}$F)
Mass			Degrees Fahrenheit ($^{\circ}$F)	5/9 x ($^{\circ}$F-32)	Degrees Celsius ($^{\circ}$C)
Ounces (oz.)	28.35	Grams (g)			
Pound (lb.)	0.45	Kilograms (kg)			
Short ton (2,000 lbs.)	0.9	Tones-metric ton (t.)			
Grams (g)	0.035	Ounces (oz.)			
Kilograms (kg)	2.2	Pounds (lbs.)			
Tonnes (t.)	1.1	Short tons (2,000 lbs.)			

Metric and Standard Measurements
Metric Conversions

Flow Rate

1 gallon per minute	$= 4.42 \times 10^{-3}$ acre feet/day
	$= 6.31 \times 10^{-5}$ m³/day
	$= 5.42$ m³/day
1 cubic foot per second	$= 449$ gallon/min. (gpm)
	$= 0.0283$ m³/sec.
	$= 2450$ m³/day
1 cubic meter per second	$= 1.58 \times 10^{4}$ gpm
	$= 35.3$ cfs
	$= 8.64 \times 10^{4}$ m³/sec.
1 cubic meter per day	$= 0.183$ gpm
	$= 4.09 \times 10^{-4}$ cfs
	$= 1.16 \times 10^{-5}$ m³/sec.
	$= 2.23 \times 10^{-3}$ cubic feet/sec. (cfs)

Velocity

1 foot/second	$= 0.682$ miles/hour
	$= 0.3048$ meters/second
1 mile/hour	$= 1.467$ feet/second
	$= 1.609$ kilometers/hour
1 meter/second	$= 3.6$ kilometers/hour
	$= 3.28$ feet/second
	$= 2.237$ miles/hour
1 kilometer/hour	$= 0.621$ miles/hour

Length

Unit	Number of Meters
Kilometer	1,000
Hectometer	100
Decameter	10
Meter	1
Decimeter	0.1
Centimeter	0.01
Millimeter	0.001

Area

Unit	Number of Square Meters
Sq. kilometer	1,000,000
Hectare	10,000
Are	100
Centare	1
Sq. centimeter	0.0001

Volume

Unit	Number of Liters
Kiloliters	1,000
Hectoliters	100
Decaliter	10
Liter	1
Deciliter	0.1
Centiliter	0.01
Milliliter	0.001

Mass

Number of Unit	Grams
Metric ton or tonne	1,000,000
Kilogram	1,000
Hectogram	100
Decagram	10
Gram	1
Decigram	0.1
Centigram	0.01
Milligram	0.001

International Metric System
Conversions

Linear	
10 millimeters (mm)	= 1 centimeter (cm)
10 cm	= 1 decimeter (dm)
10 dm	= 1 meter (m)
1,000 m	= 1 kilometer

Weight	
10 decigrams	= 1 gram
1,000 grams	= 1 kilogram
1,000 (kg)	= 1 metric ton

Area	
10,000 square meters	= 1 hectare
100 hectares	= 1 square kilometer

Liquid	
1,000 milliliters (ml)	= 1 liter

Water Quality Unit Conversions

1 kilogram (kg)	= 1000 grams (g)
1 gram (g)	= 1000 milligrams (mg)
1 kilogram (kg)	= 1 million milligrams (mg)
1 kilogram (kg)	= 1 billion micrograms (μg)
1 milligram (mg)	= 1000 micrograms (μg)

Concentrations:

1 gram/L	= 1 part per thousand (ppt)
1 milligram/L	= 1 part per million (ppm)
1 microgram/L	= 1 part per billion (ppb)

Comparison of International Drinking Water Guidelines[1]

Parameter	USEPA[2] Maximum Contaminant Level (MCL)	Canada[3] Maximum Acceptable Concentration	EEC[4] Maximum Admissible Concentration	Japan[5] Maximum Admissible Concentration	WHO[6] Guideline	Bottled Water U.S. Federal Drug Administration
Aluminum Ammonium Antimony Arsenic	0.05-0.2 mg/L[7] 0.006 mg/L 0.05 mg/L	 0.025 mg/L	0.2 mg/L 0.5 mg/L 0.01 mg/L 0.05 mg/L	0.2 mg/L No standard 0.002mg/L[8] 0.01 mg/L	0.2 mg/L 1.5 mg/L 0.005 mg/L 0.01 mg/L	 0.05 mg/L
Barium Boron Cadmium Chloride	2.0 mg/L 0.005 mg/L 250mg/L[7]	1.0 mg/L 5.0 mg/L 0.005 mg/L 250 mg/L	No standard 1.0 mg/L 0.005 mg/L 250 mg/L	No standard 0.2 mg/L[8] 0.01 mg/L 200 mg/L	0.7 mg/L 0.3 mg/L 0.003 mg/L 250 mg/L	2.0 mg/L 0.005 mg/L
Chromium Coliforms, total Organisms/100mL Coliforms (E.Coli) Organisms/100mL Color	0.1 mg/L % positive 0 15 cu[7]	0.05 mg/L 0 0 15 cu	0.05 mg/L 0 or MPN 0 20 mg Pt-Co/L	0.05 mg/L 0 0 5 cu	0.05 mg/L 0 0 15 cu	0.1 mg/L MF <15 cu
Copper Cyanide Fluoride Hardness	1.3 mg/L[7] 0.2mg/L 2.0-4.0 mg/L[7]	1.0 mg/L 0.2 mg/L 1.5 mg/L	2.0 mg/L 0.05 mg/L 0.7-1.5 mg/L 50 mg/L	1.0 mg/L 0.01 mg/L 0.8 mg/L 300 mg/L	1-2 mg/L 0.07 mg/L 1.5 mg/L	1.0 mg/L
Iron Lead Manganese Mercury	0.3 mg/L[7] 0.015 mg/L 0.05 mg/L 0.002 mg/L	0.3 mg/L 0.01 mg/L 0.05 mg/L 0.001 mg/L	0.2 mg/L 0.01 mg/L 0.05 mg/L 0.001 mg/L	0.3 mg/L 0.05 mg/L 0.01-0.05 mg/L 0.0005 mg/L	0.3 mg/L 0.01 mg/L 0.1-0.5 mg/L 0.001 mg/L	 0.005 mg/L 0.002 mg/L
Molybdenum Nickel Nitrate/Nitrite, total Nitrate	 0.1 mg/L 10.0 mg/L as N 10.0 mg/L as N	 10.0 mg/L as N	 0.02 mg/L 50 mg/L	0.07 mg/L 0.01 mg/L[8] 10.0 mg/L as N 10 mg/L as N	0.07 mg/L 0.02 mg/L 50 mg/L as NO$_3$	 10 mg/L as N
Nitrite Odor pH Phosphorus	1 mg/L as N 3 TON[9] 6.5-8.5	3.2 mg/L 6.5-8.5	0.1 mg/L 2 dilution no. @12°C; 3 dilution no. @ 25°C. 6.5-9.5 5 mg/L	10 mg/L 3 Ton 5.8-8.6 No standard	3 mg/L as NO$_2$ 6.5-8.5	1 mg/L as N
Phenols Potassium Selenium Silica Dioxide	 0.05 mg/L	0.002 mg/L 0.01mg/L	0.5 µg/L C$_6$H$_5$OH 12 mg/L 0.01 mg/L 10 mg/L	0.005 mg/L No standard 0.01 mg/L No standard	 0.01 mg/L	 0.05 mg/L
Silver	0.1mg/L[7]	0.05 mg/L	0.01 mg/L	No standard	No standard	
Solids, total dissolved Sodium Sulfate	500 mg/L[7] 250mg/L[7]	500 mg/L 500 mg/L	No standard 75-150 mg/L 250 mg/L	500 mg/L 200 mg/L No standard	1000 mg/L 200 mg/L 250 mg/L	
Turbidity Zinc	0.5-5 NTU 5 mg/L[7]	1 NTU 5.0 mg/L	4 JTU No standard	1-2 units 1.0 mg/L	5 NTU 3.0 mg/L	<5 NTU

Comparison of International Drinking Water Guidelines

1   To our knowledge, data in this table were accurate and current at the publication date. Contact the regulatory agency in your area for the most current information.

2   United States Environmental Protection agency.

3   These limits are established by Health Canada.

4   In the EEC (European Economic Community), these limits are set by the European Committee for Environmental Legislation.

5   In Japan, these limits are established by the Ministry of Health and Welfare.

6   World Health Organization.

7   U.S. Secondary MCL.

8   Identified as a parameter to be regulated in the future.

9   Threshold Odor Number.

Reprinted with permission from the Hach Company, Loveland, CO. Hach Company. 2002. Water Analysis Handbook; 4th Edition. Loveland, CO: Hach Company.

Milestones
in Water Quality Management

From the beginning, humans have been challenged with finding and providing clean, healthy water to drink. Through the millennia there have been many contributions to the science and management of water quality, both for human consumption as well as in the environment. This list provides a small sample of the contributions and milestones in the history of water quality management.

4000 B.C. Ancient Sanskrit and Greek writings recommend water treatments like filtering through charcoal, exposing to sunlight, boiling, and straining

1500 B.C. Egyptians use alum to settle out suspended particles in water

400 B.C. Hippocrates states importance of water quality to health, recommends boiling and straining water

312 B.C. Start of Roman aqueduct construction

144 B.C. Aqua Marcis, the longest Roman aqueduct built

1652 A.D. First incorporated waterworks form in Boston

1700s Filtration found effective for removing particles suspended in water

1774 Chlorine is discovered in Sweden

1800s Slow sand filtration used in Europe

1804 The first municipal water filtration works opens in Paisley, Scotland

1835 Chlorine is first applied to drinking water to control foul odors in the water

1849 Cholera epidemic—8,000 lives claimed in New York City and 5,000 in New Orleans

1850 Swamp Act—Encourages draining of wetlands for development

1854 Dr. John Snow discovers cholera outbreak in London is due to a contaminated well on Broad Street

1862 Homestead Act—Opens the West to settlement and water development

1877 Louis Pasteur develops the theory that germs spread disease

1880s Louis Pasteur demonstrated the "germ theory" of disease—how microbes transmit disease through water

1882 Filtration of London drinking water begins

1890s Chlorine is proven an effective disinfectant of drinking water

1890 National Weather Service—Collects data to monitor weather patterns, monitors weather data

1894 Carey Irrigation Act—Grants public lands to states for irrigation

1896 Louisville Water Company creates new treatment technique by combining coagulation with rapid-sand filtration

1899 Rivers and Harbors Act—Prohibits discharge of solids into navigable rivers

1900s	U.S. drinking water treatment systems built to reduce turbidity and the microbial contaminants associated with sediment; use slow sand filtration
1902	Belgium implements the first continuous use of chlorine to make drinking water biologically "safe"
1906	General Dam Act—Regulates private dam construction on navigable streams
1908	U.S. public water supply is chlorinated for the first time at Boonton reservoir supply, Jersey City, NJ
1912	Congress passes the Public Health Service Act, authorizes surveys and studies for water pollution
1914	First standards under the Public Health Service Act become law, establish maximum contaminant limit for drinking water
1924	Oil Pollution Act—Prohibits the discharge of oil into marine waters
1925	Rivers and Harbors Act—Authorizes the U.S. Army Corps of Engineers to survey all navigable waters and formulate general water use plans
1925	Public Health Service—Revises standards for drinking water
1936	Flood Control Act—First nationwide flood control act; introduces cost-benefit analysis
1944	Flood Control Act—Recognizes the priority of flood control over irrigation, recreation, and power production
1946	Public Health Service—Again revises standards for drinking water
1948	Water Pollution Control Act—Provides technical assistance to municipalities for wastewater treatment

1954	Small Watershed Act—Sets up a small watershed program under the Soil Conservation Service (now Natural Resources Conservation Service)
1955	An infectious hepatitis epidemic in New Delhi, India is traced to inadequately chlorinated water at a treatment plant; 1 million people infected
1956	Federal Water Pollution Control Act—Increases federal assistance for wastewater treatment
1958	Fish and Wildlife Coordination Act—Requires equal consideration of wildlife protection at federal water projects
1962	U.S. Public Health Service Drinking Water Standards Revision is accepted as minimum standards for all public water suppliers
1962	Water Resource Research Act—Establishes water resource research institutes in each state and territory with a $100,000 grant
1965	Water Resource Planning Act—Establishes the Federal Water Pollution Control Commission
1960s (late)	Industrial and agricultural advances and creation of man-made chemicals have negative impacts on environment and public health
1969	U.S. Public Health Service Community Water Supply study reveals major deficiencies in the nation's public water supplies
1969	Izaak Walton League—Establishes SOS (Save Our Streams) program; river and stream monitoring
1971-78	Maine, Minnesota, Michigan, and New Hampshire—Initiate statewide lake monitoring programs

1972	Clean Water Act—Amendment to the Federal Water Pollution Control Act	

1972 Federal Water Pollution Control Act Amendments–Institutes a national permit system for point source discharges; puts U.S. Army Corps of Engineers in charge of regulating discharge of dredge and fill materials

1973 U.S. Congress debates new legislative proposals for federal safe drinking water law due to studies on the Mississippi River

1973 Endangered Species Act—Protects endangered species

1974 Safe Drinking Water Act—Coordinates monitoring and training for safe drinking water; sets up drinking-water standards

1977 Safe Drinking Water Act is amended to extend authorization for technical assistance information, training, and grants to the states

1977 Clean Water Act amendment—Authorizes more grant money for states

1980s Improvements made in membrane development for reverse osmosis filtration and ozonation

1981 Clean Water Act amendment—Authorizes more grant money for states

1985 Food Security Act—Establishes erosion control programs for agricultural lands; denies federal farm benefits to farmers harvesting from converted wetlands

1985 Rhode Island and states surrounding the Chesapeake Bay—Initiate estuary monitoring

1986 Safe Drinking Water Act—Further amended to set mandatory deadlines for the regulation of key contaminants

1987 Wild and Scenic Rivers Act—Protects instream flows for rivers designated wild and scenic

1987 Clean Water Act amendment—Authorizes more grant money for states

1987 Water Quality Act—Requires EPA to regulate storm water runoff and states to prepare nonpoint source management programs

1988 First National Volunteer Monitoring Conference—85 participants

1989 First Issue of *The Volunteer Monitor*—8 pages, 3000 copies distributed

1994 Volunteer Water Monitors—517 Programs in 45 states; *The Volunteer Monitor*: 24 pages, 20,000 copies distributed

1996 Safe Drinking Water Act—Reauthorized

1998 Volunteer Water Monitors—800+ programs in all 50 states; *The Volunteer Monitor*: 24 pages, 40,000 copies distributed

2000 Centers for Disease Control and Prevention and the National Academy of Engineering–Named water treatment as one of the most significant public health advancements of the 20th Century

2001 More than 90% of the U.S. population is served by community water systems

References:

National Drinking Water Clearinghouse. 2001. *On Tap.* Retrieved February 15, 2002 from www.ndws.wvu.edu

The Watercourse. 1995. *Project WET Curriculum and Activity Guide.* Bozeman, MT: The Watercourse.

United States Environmental Protection Agency. *The History of Drinking Water Treatment.* February 2000. Retrieved February 15, 2002 from http://www.epa.gov/safewater/sdwa/trends.html

Weise, J. No date available. *Historic Milestones in Drinking Water History.* Retrieved February 16, 2002 from http://www.state.ak.us/dec/deh/water/history.htm

Glossary

accuracy. A measure of how close results are to the actual value.

acid. A substance that has a pH of less than 7, which is neutral. Specifically, an acid has more free hydrogen ions (H^+) than hydroxide ions (OH^-).

aeration. The mixing or turbulent exposure of water to air and oxygen.

alkalinity. The capacity of water for neutralizing an acid solution.

ammonia. A gaseous compound of hydrogen and nitrogen, NH_3, with a pungent smell and taste; often called volatile alkali.

amoeba. A protozoan of the genus *Amoeba* or related genera, occurring in water and soil and as a parasite in other animals.

aquifer. A geologic formation(s) that is water bearing. A geological formation or structure that stores and/or transmits water, such as to wells and springs.

assessment. The act of assessing; appraisal; monitoring important parameters

bacteria/bacterium. Microscopic, single celled organisms; some of which contribute to decomposition processes noted in wetland environments, others may cause serious illness if consumed or contacted.

base. A substance that has a pH of more than 7, which is neutral. A base has more free hydroxide ions (OH^-) than hydrogen ions (H^+).

benthic. Referring to organisms that live on the bottom of water bodies.

Best Management Practices (BMP). Methods determined by land and water managers to describe land use measures designed to reduce or eliminate nonpoint source pollution.

bioassessment. The use of biological surveys and other direct measurements of living systems within a watershed.

biodiversity. A measure of the different species of organisms in a defined area; a measure of biological diversity.

buffer. A solution or liquid whose chemical makeup neutralizes acids or bases without a great change in pH. Surface waters and soils with chemical buffers are not as susceptible to acid deposition as those with poor buffering capacity. (See alkalinity.)

coagulation. In water treatment, the use of chemicals to make suspended solids gather or group together into small flocs.

colloid. Finely divided solids which will not settle but which may be removed by coagulation or biochemical action.

comparability. How well water quality data compares with data from the same project or other projects. Comparability is maximized when standard protocols are used to collect data.

completeness. A comparison of the minimum number of samples necessary for meaningful water quality data analysis and the number intended to be collected. It is wise to collect extra samples in case some are declared unusable.

concentration (of a solution). Amount of a chemical or pollutant in a particular volume of water.

concept map. A structured process, focused on a topic of interest, involving input from one or more participants that produces an interpretable pictorial view of their ideas and concepts and how these are interrelated.

condensation. The process by which a gas or vapor changes to a liquid or solid; also the liquid or solid so formed. The opposite of evaporation. In meteorological usage, this term is applied only to the transformation from vapor to liquid.

confluence. The flowing together of two or more streams. A point of juncture.

contaminant. Any substance that when added to water (or another substance) makes it impure and unfit for consumption or use.

criteria. Water quality conditions which are to be met in order to support and protect desired uses.

cryptosporidium. A protozoan of the genus cryptosporidium that is an intestinal parasite in humans and other vertebrates and some times causes diarrhea that is especially severe in immuno-compromised (weakened immune system) individuals.

cumulative. To increase by succes-sive additions; as, a cumulative pollutant whose toxicity increases with subsequent additions.

downriver. Toward or near the mouth of a river; in the direction of the current.

disinfection. To cleanse so as to destroy or prevent the growth of disease-carrying microorganisms.

dissolved oxygen (DO). Amount of oxygen gas dissolved in a given quantity of water at a given temperature and atmospheric pressure. It is usually expressed as a concentration in parts per million or as a percentage of saturation.

Drinking Water Standards. In the U.S. National Primary Drinking Water Regulations (NPDWRs or primary standards) are legally enforceable standards that apply to public water systems. Primary standards protect health by limiting the levels of contaminants in drinking water.

emulsion. A suspension of small globules of one liquid in a second liquid with which the first will not mix.

epidemiologist. A person who studies the incidence, transmis-sion, distribution, and control of infectious disease (including waterborne disease) in large populations.

erosion. The process in which a material is worn away by a stream of liquid (water) often due to the presence of abrasive particles in the stream.

estuary. The part of the lower wide course of a river where its current is met by the tides, typically containing a mixture of fresh and salt water (brackish).

eutrophication. Having water rich in mineral and organic nutrients that promote a proliferation of plant life, especially algae, which overproduce, die off, and the bacteria that decomposes them eventually reduces the dissolved oxygen content, sometimes causing the extinction of other organisms (fish, macro-invertebrates); typically in a lake or pond.

evaluate. To ascertain or fix the value or worth of. To examine and judge carefully; appraise.

evaporation. (1) The physical process by which a liquid (or a solid) is transformed to the gas-eous state. (2) Water from land areas, bodies of water, and all other moist surfaces is absorbed into the atmosphere as a vapor.

exponential growth (statistics). A rate of growth characterized by a fixed percentage each time period, e.g., a ten percent growth each period of time. Represented by the equation: $N(t) = N_0 e^{kt}$, where $N(t)$ is the population level at time t, e is the base of the natural logarithm, k is a constant value, and t represents time period where $t = 1,2,...n$.

extremophile. Microbe that lives in environments that are usually unfit for life, including those of high salinity, very low or very high pH, and very low or very high tem-peratures.

falls (waterfall). A steep descent of water from a height; a cascade.

filtration. The mechanical process which removes particulate matter by separating water from solid material, usually by passing it through sand.

floc. Generally, a very fine, fluffy mass formed by the aggregation of fine suspended particles, as in a precipitate. In terms of water quality, clumped solids or precipitates formed in sewage by biological or chemical activity. Clumps of bacteria and particulate impurities that have come together and formed a cluster. Found in flocculation tanks and settling or sedimentation basins.

giardiasis. A disease that results from an infection by the protozoan parasite *Giardia intestinalis* (Giardia flagellates) caused by drinking water that is either not filtered or not chlorinated. This disorder is more prevalent in children than in adults and is characterized by abdominal discomfort, nausea, and alternating constipation and diarrhea.

ground water. (1) Water that flows or seeps downward and saturates soil or rock, supplying springs and wells. The upper surface of the saturate zone is called the water table. (2) Water stored underground in rock crevices and in the pores of geologic materials that make up Earth's crust.

headwaters. The water from which a river rises; a source.

heterogeneous. Non-uniform in structure or composition throughout.

homogeneous. Uniform in structure or composition throughout.

hydrogen ion (H$^+$). The positively charged ion of hydrogen, H$^+$, formed by removal of the electron from atomic hydrogen and found in all aqueous solutions of acids.

hydrologic cycle. The cyclic transfer of water vapor from the Earth's surface via evapotranspiration into the atmosphere, from the atmosphere via precipitation back to Earth, and through runoff into streams, rivers, and lakes, and ultimately into the oceans.

hydroxide ion (OH$^-$). The anion $^-$OH, also called hydroxyl ion.

impermeable. Material that does not permit liquids to pass through.

indicator (ecology). A quantitative or qualitative variable which can be measured or described and which when observed periodically demonstrates trends. Ecosystem indicators track the magnitude of stress, habitat characteristics, exposure to the stressor, or ecological response to exposure.

Internet. An interconnected system of networks that connects computers around the world.

ion. (1) An element or compound that has gained or lost an electron, so that it is no longer neutral electrically, but carries a charge. (2) An atom or molecule that carries a net charge (either positive or negative) because of an imbalance between the number of protons and the number of electrons present. If the ion has more electrons than protons, it has a negative charge and is called an anion; if it has more protons than electrons it has a positive charge and is called a cation. (3) (water quality) An electrically charged atom that can be drawn from waste water during electrodialysis.

logarithm The value of the exponent that a fixed number (the base) must have to equal a given number. It is calculated as $b^x = y$, where b is the base and x is the logarithm. The base for the common logarithm is 10. As an example, the logarithm of 100 is 2 since 10^2 is equal to 100. This may also be written as $\log_{10} 100 = 2$. The base of the Natural Logarithm is approximately equal to 2.718282.

macroinvertebrate. Invertebrate animal (without back bones) large enough to be observed without the aid of a microscope or other magnification.

marine. Relating to seawater or saline waters that are part of an ocean or sea.

Maximum Contaminant Level (MCL). The designation given by the U.S. Environmental Protection Agency (EPA) to water-quality standards promulgated under the Safe Drinking Water Act. The MCL is the greatest amount of a contaminant that can be present in drinking water without causing a risk to human health.

Methyl Tertiary Butyl Ether (MTBE). A oxygenate and gasoline additive used to improve the efficiency of combustion engines in order to enhance air quality and meet air pollution needs. MTBE has been found to mix and move more easily in water than many other fuel components, thereby making it harder to control, particularly once it has entered surface or ground waters.

microbe. Short for microorganism. Small organisms that can be seen only with the aid of a microscope. The term encompasses viruses, bacteria, yeast, molds, protozoa, and small algae; however, microbe is used more frequently to refer to bacteria. Microbes are important in the degradation and decomposition of organic materials added to the environment by natural and artificial mechanisms.

micrograms per liter (μg/l). Micrograms per liter of water.

One thousand micrograms per liter is equivalent to 1 milligram per liter. This measure is equivalent to parts per billion (ppb).

midriver. The middle or center of a river.

mixture. Compounds containing other substances. For example, water that is found in nature is a mixture because it contains other substances dissolved in it.

neutralize. The equalization of hydrogen and hydroxyl ion concentrations such that the resulting solution is neither acidic nor basic; also, decreasing the acidity or alkalinity of a substance by adding alkaline or acidic materials, respectively.

nitrate. A salt of nitric acid; a common water pollutant.

nonpoint source pollution (NPS). Pollution discharged over a wide land area, not from one specific location. Nonpoint source pollution is contamination that occurs when rainwater, snowmelt, or irrigation washes off plowed fields, city streets, or suburban backyards. As this runoff moves across land surface, it picks up soil particles and pollutants, such as nutrients and pesticides.

nutrient. As a pollutant, any element or compound, such as phosphates or nitrates, that fuels abnormally high organic growth in aquatic ecosystems.

oxygenate. To treat, combine, or infuse with oxygen.

parts per million (ppm). Units typically used in measuring the number of "parts" by weight of a substance in water; commonly used in representing pollutant concentrations. Equal to milligrams per liter (mg/l).

parts per billion (ppb). Units typically used in measuring the number of "parts" by weight of a substance in water; commonly used in representing pollutant concentrations. Equal to micrograms per liter (μg/l).

pathogen. Microorganism which can cause disease.

percolation. (1) The slow seepage of water into and through the ground. (2) The slow seepage of water through a filter medium. (3) The movement, under hydrostatic pressure, of water through the interstices of a rock or soil.

pH. A measure of the relative acidity or alkalinity of water. Water with a pH of 7 is neutral; lower pH levels indicate increasing acidity, while pH levels higher than 7 indicate increasingly basic solutions.

phosphate. Salt of phosphoric acid, often found in fertilizers.

point source pollution. Water pollution coming from a single point, such as a sewage-outflow pipe.

potable. Fit to drink.

precision. The repeatability of a series of tests results; when the testing method gives the same answer under the same set of circumstances or sampling criteria.

Primary Drinking Water Standard. Enforceable standards related directly to the safety of drinking water; set by the U.S. Environmental Protection Agency (EPA).

protozoa. A phylum or subkingdom including all single-celled animals with membrane bound organelles; they may be aquatic or parasitic, with or without a test, solitary or colonial, sessile or free-swimming, moving by cilia, flagella, or pseudopodia. (From a Greek phrase meaning first animal.)

purification (water). Steps taken to eliminate impurities and pollution from water.

remediate (environmental). Cleanup or other methods used to remove or contain a toxic spill or hazardous materials from a Superfund site or other impaired site or waterway.

replicate. To duplicate, copy, reproduce, or repeat, as in replicate water quality samples.

representativeness (statistics). How well a given sample represents the total population from which it was taken.

restoration. The process of bringing back to existence, or reestablishing, the original condition of a degraded environment.

runoff. An overland flow of water entering rivers, freshwater lakes, or reservoirs. Contaminants are often picked up by runoff and carried to the receiving water body. Primary source of nonpoint source pollution. (See stormwater runoff.)

salinity. The relative proportion of salt in a solution, typically measured in g/l, or parts per thousand (ppt). Seawater is typically around 35 ppt.

saturated zone. In the saturated zone, every available space between soil particles and in rock formations is filled with water.

Secondary Drinking Water Standard. Non-enforceable standards related to the aesthetic quality of drinking water such as those relating to taste and odor; generally set by the U.S. Environmental Protection Agency (EPA) or state water-quality enforcement agencies based on EPA guidance.

sediment. (1) Material that settles to the bottom of a liquid. (2) Solid fragments of inorganic material that come from the weathering of rock and are carried and deposited by wind, water, or ice.

sedimentation. (1) A large-scale water treatment process where heavy solids settle out to the bottom of the treatment tank after flocculation. (2) The act of depositing a sediment; specifically (Geol.), the deposition of the material of which sedimentary rocks are formed.

sludge. Semisolid material such as the type precipitated by sewage treatment. Mud, mire, or ooze covering the ground or forming a deposit, as on a riverbed.

solute. A substance that is dissolved in another substance, thus forming a solution.

solution. A mixture of a solvent and a solute. In some solutions, such as sugar water, the substances mix so thoroughly that the solute cannot be seen. But, in other solutions, such as water mixed with dye, the solution is visibly changed.

solvent. A substance that dissolves other substances, thus forming a solution. Water dissolves more substances than any other, and is known as the "universal solvent."

source water. The point at which water springs into being or from which it derives or is obtained. The point of origin, such as a spring, of a stream or river.

standards. An acknowledged measure of comparison for quantitative or qualitative value; a criterion. An object that under specified conditions defines, represents, or records the magnitude of a unit.

stormwater runoff. The water and associated material draining into streams, lakes, or sewers as the result of a storm.

sublimation. (1) The transformation of a solid to the gaseous phase without passing through the normally intermediate liquid phase. (2) The change of a solid to a vapor (or the reverse) without the appearance of a liquid state, as in the changing of snow directly into water vapor without melting.

surface area (lake or impoundment). The extent of a 2-dimensional surface enclosed within a boundary.

surface water. Water that is on Earth's surface, such as in a stream, river, lake, or reservoir. A natural or artificial pond or lake used for the storage and regulation of water.

suspension. The state in which the particles of a substance are mixed with a liquid but are not dissolved.

toxicity. (1) The ability of a chemical substance to cause acute or chronic adverse health effects in animals, plants, or humans when swallowed, inhaled or absorbed. (2) The occurrence of lethal or sub-lethal adverse effects on representative, sensitive organisms due to exposure to toxic materials.

transpiration. (1) The movement of water from the soil or ground water reservoir via the stomata in plant cells to the atmosphere. (2) The process by which water vapor escapes from a living plant, principally through the leaves, and enters the atmosphere. Transpiration, combined with evaporation from the soil, is referred to as evapotranspiration.

turbidity. The amount of solid particles that are suspended in water and that cause light rays shining through the water to scatter. Thus, turbidity makes the water cloudy or even opaque in extreme cases. Turbidity is measured in nephelometric turbidity units (NTU).

Tyndall effect. Visible pattern caused by reflection of light from suspended particles in a solution or the atmosphere.

unicellular. Having, or consisting of, a single cell, as in a unicellular organism.

unsaturated zone. The zone immediately below the land surface where the pores contain both water and air, but are not totally saturated with water. These zones differ from an aquifer, where the pores are saturated with water.

virus. An organism that causes and transmits an infectious disease.

water table. The uppermost level of water in the saturated part of an aquifer.

water treatment. Systematic purification of water for human consumption.

waterborne disease. Any illness transmitted through ingestion of or contact with water contaminated by disease-causing organisms (e.g., bacteria, viruses, or protozoa) or chemicals.

watershed. The land area that drains water into a particular stream, river, or lake. It is a land feature that can be identified by tracing a line along the highest elevations between two areas on a map, often a ridge. Large watersheds, like the Mississippi River basin, contain thousands of smaller watersheds.

Index